BENGAL DISTRICT GAZETTEERS

CUTTACK

BENGAL DISTRICT GAZETTEERS

Burdwan/ **J.C.K. Paterson**
Darjeeling/ **L.S.S. O'Malley**
Howrah/ **L.S.S. O'Malley**
Hooghly/ **L.S.S. O'Malley**
Santal Parganas/ **L.S.S. O'Malley**
Sambalpur/ **L.S.S. O'Malley**
Puri/ **L.S.S. O'Malley**
Gaya/ **L.S.S. O'Malley**
24 - Parganas/ **L.S.S. O'Malley**
Purnea/ **L.S.S. O'Malley**
Bhagalpur/ **J. BYRNE**
Feudatory States of Orissa/ **L.E.B. Cobden-Ramsay**
Sinhbhum, Saraikela and Kharsawan/ **L.S.S. O'Malley**
Cuttack/ **L.S.S. O'Malley**
Darbhanga/ **L.S.S. O'Malley**
Champaran/ **L.S.S. O'Malley**

BIHAR AND ORISSA DISTRICT GAZETTEERS

Patna/ **L.S.S. O'Malley**
Shahabad/ **L.S.S. O'Malley**
Monghyr/ **L.S.S. O'Malley**
Saran/ **L.S.S. O'Malley**

EASTERN BENGAL DISTRICT GAZETTEERS

Dacca/ **B.C. Allen**

BENGAL DISTRICT GAZETTEERS

CUTTACK

L. S. S. O'MALLEY,

LOGOS PRESS
New Delhi-110 002

ISBN-13: 978-81-7268-229-3

First Published 1906
Reprinted 2013

Printed and Published by
LOGOS PRESS

Building No. 4788-90, Street No. 23, Ansari Road, Darya Ganj,
New Delhi-110 002 (India)
Phone : 23272187
Email : mail@logospressindia.com
Website : www.logospressindia.com

PREFACE

THE district of Cuttack possesses a more complete bibliography than most of the districts of Bengal. A full account of the people and their circumstances was written by Mr. Stirling in the beginning of the nineteenth century, and his Account of Orissa Proper or Cuttack, published in 1822, is still the principal authority for the early history of the district under British rule. The modern authority is Mr. Maddox's Final Report on the Survey and Settlement of the Province of Orissa, 1890 to 1900 A.D., which contains an exhaustive review of the economic condition of that Province; and the intermediate authorities are Toynbee's Sketch of the History of Orissa from 1803 to 1828, published in 1873, and Sir W. W. Hunter's account published in 1877 in vol. XVIII of the Statistical Account of Bengal. In writing the present volume I have made use of all these sources of information, and especially of Mr. Maddox's Report, which has been described as "a veritable Encyclopaedia of the Province." My thanks are due to Babu Jamini Mohan Das, M.A., B.L., for his ready assistance in revising the description of the religions and castes of Cuttack and for a note on the Utkal Brahmans ; and I am especially indebted to Babu Monmohan Chakravarti, M.A., B.L., M.R.A.S., who has been so good as to write for the Gazetteer an account of the history of Orissa up to the time of the Muhammadan invasion.

L. S. S. O'MALLEY

PREFACE

PLAN OF CONTENTS

TABLE OF CONTENTS

CHAPTER I

PHYSICAL ASPECTS

CHAPTER II

HISTORY

CHAPTER III

THE PEOPLE

CHAPTER IV

PUBLIC HEALTH

CHAPTER V

AGRICULTURE

CHAPTER VI

EMBANKMENTS AND CANALS

CHAPTER VII

NATURAL CALAMITIES

CHAPTER VIII

RENTS, WAGES AND PRICES

CHAPTER IX

OCCUPATIONS, MANUFACTURES AND TRADE

CHAPTER X

MEANS OF COMMUNICATION

CHAPTER XI

LAND REVENUE ADMINISTRATION

CHAPTER XII

GENERAL ADMINISTRATION

CHAPTER XIII

LOCAL SELF-GOVERNMENT

CHAPTER XIV

EDUCATION

CHAPTER XV

GAZETTEER

GAZETTEER

OF THE

CUTTACK DISTRICT.

CHAPTER I.

PHYSICAL ASPECTS.

CUTTACK, the central district of the Orissa Commissionership or Division, is situated between 20° 2′ and 21° 10′ north latitude, and between 85° 21′ and 87° 1′ east longitude. It contains an area, according to the latest survey, of 3,654* square miles; and a total population, as ascertained by the census of 1901, of 2,062,758 souls. The principal town, which is also the administrative head-quarters of the district and of the Orissa Division, is Cuttack, situated on a tongue of land formed by the Mahānadī and Kātjurī rivers at their point of bifurcation, in 20° 29′ north latitude and 85° 52′ east longitude. GENERAL DESCRIPTION.

The town was formerly one of the royal strongholds of ancient Orissa, and still contains the remains of the citadel in which its rulers once held their court. From this circumstance it derived the name of *Kataka* or fort; and, as is frequently the case, the designation of the capital town was in course of time given to the surrounding country. Origin of name.

The district is bounded on the north by the Baitaranī river and Dhāmra estuary, which separate it from the district of Balasore; on the east by the Bay of Bengal; on the south by the district of Purī; and on the west by the Tributary States of Orissa. Boundaries.

It consists of three distinct tracts differing widely in their physical aspects. The first is a marshy woodland strip, from 3 to 30 miles in breadth, extending along the coast from the river Dhāmra on the north to the Devī on the south; the second is a Natural divisions.

* The area of the district shewn in the census report of 1901 was 3,624 square miles; the area shewn above is that reported by the Surveyor-General.

B

cultivated alluvial plain formed from the deposits of its great rivers; and the third is a broken hilly region forming the western boundary of the district.

The marshy strip along the coast is a low woodland tract, abounding in swamps and morasses and intersected by innumerable winding creeks with a coarse jungly growth of canes, brushwood and reedy grass on either side. It has aptly been described as the Sundarbans on a miniature scale, and it resembles that tract in its swamps, dense jungle and noxious atmosphere. This dismal region is subject to inundations of sea water, which leave a deposit of salt on the surface of the low-lying country, and in many parts render cultivation impossible. The only means of communication are the small sluggish streams winding into the heart of the jungle, along which country boats convey supplies of wood to the villages in the adjoining delta. The latter consists of a level plain stretching inland for about 40 miles, and occupying the country between the marshy sea-coast strip and the hilly frontier. It is intersected by several large rivers, which emerge from the western mountains and throw out a network of branches in every direction; these, after innumerable twists and interlacings, frequently rejoin the parent stream as it approaches the sea. It is a region of rich rice-fields, dotted with magnificent banyan trees, thickets of bamboos, mango orchards and palm groves of exquisite foliage; and it forms the only really fertile part of the district.

HILLS. The frontier separating the district from the Tributary States on the west consists of a chain of hills, covering about 46 square miles, with thickly wooded slopes and fertile valleys between. The greatest distance of this hilly region from the sea coast is about 60 to 70 miles, but in many places the breadth of the alluvial plain does not exceed 15 to 20 miles. The hills do not consist of long continuous ranges, but are generally found in irregularly scattered groups, running nearly due east and west for a distance of about 15 miles. With the exception of a few naked bluffs, they are for the most part covered with vegetation; their outline, however abrupt, is always more or less rounded; and it is evident that they owe their present form to marine action. On the other hand, some hills, which appear from a distance to be flat-topped, really consist of a series of steep rugged ridges separated by deep precipitous valleys cut out by the denuding action of running water. In this western tract lie all the hills of the district with the exception of a few isolated peaks which break the evenness of the plain to the north of Cuttack. None of them are more than 2,500 feet high, but many are of great interest on account of the shrines or ancient forts with which they are crowned. The most interesting hills are in the Assia range, particularly Naltigiri, with its sandal trees and Buddhist remains; Udayagiri, with its colossal image of Buddha, sacred reservoir, ruined temples and caves; Assiagiri, the highest hill in the district, standing 2,500 feet above the sea, with an old mosque nearly 200 years old; and the Mahāvināyaka peak in *Kilā* Darpan, which has been consecrated for ages to Siva-worship by the devout ascetics and pious pilgrims who have penetrated its dense jungle. This hilly borderland and the low lands along the coast were formerly known as the Rājwāra or Zamīndāra, and were held by feudal chiefs, who paid a tribute to their overlord, but otherwise retained an independent power; while the wide alluvial plains forming the delta of the Mahānadī, Brāhmanī and Baitaranī rivers constituted the Mughalbandi or Khālsa, *i.e.*, the crown lands from which the Mughal conquerors, like the indigenous sovereigns before them, derived the greater part of their revenue.

The most conspicuous feature in the general aspect of the district is its system of rivers, which issue in three magnificent streams through three great gorges in the mountainous country to the west. To the south, the Mahānadī debouches upon the plains just above Narāj, 70 miles from the sea; on the extreme north of the district, the sacred Baitaranī emerges from a more open country, and forms the boundary between Cuttack and Balasore; and the Brāhmanī enters the district about halfway between the two. Cuttack is thus divided into two great valleys, one of them lying between the Baitaranī and Brāhmanī and the other between the Brāhmanī and Mahānadī. RIVER-SYSTEM.

During the hot weather the upper channels of these rivers dwindle to insignificant streams dotted here and there with stagnant pools; but in the rainy season they bring down an enormous mass of water from the high table-lands in which they take their rise. Towards the coast they gradually converge, and pouring down their accumulated waters upon the level plain within 30 miles of each other, are only prevented from bursting over their banks and sweeping across the country by a great system of embankments. These great rivers drain an immense area amounting to over 65,000 square miles, and the rapidity of the current acquired among the mountains brings down a vast quantity of silt in suspension. As soon however as the river reaches the plains and leaves the broken hilly region for the level delta, its current is checked. The further it goes the more sluggish does its stream become, and the river, being unable to carry down the sand with which it is charged, deposits it in its bed and on its banks. By degrees, therefore, the bed is raised, and the river flows at a higher level than the surrounding country; and the central portion rising more rapidly than the banks, the channel of the river becomes gradually shallower. The distributaries of the main rivers have their beds raised in the same way; and the result is that the rivers and their various channels become less and less able to carry off the water-supply to the sea, and frequently prove inadequate to furnish an outlet for the vast volume of water poured in at their heads during the rainy season. The velocity which these great rivers obtain in descending from the interior table-land being thus checked, they break up into a hundred distributaries radiating across the level plains. The distributaries, struggling by a thousand contortions and convolutions towards the coast, form a network of rivers, which joining here and separating there generally reunite with one of the three parent channels as they approach the sea.

The following table illustrates the main points in the river system of the district:—

BAITARANI { Baitaranī . . . . Baitaranī . . . . } Dhāmra.
BAITARANI { Burha . . . . } Kharsuā
BRAHMANI { Kharsuā . . . . { Patiyā . } Kharsuā . . . . . } Brāhmanī } Dhāmra.
BRAHMANI { Kharsuā . . . . { Kharsuā . } Kharsuā
BRAHMANI { Brāhmanī . . . . { Brāhmanī . Brāhmanī . . . . } Brāhmanī
BRAHMANI { Brāhmanī . . . . { Kimiriā . } Kelo . . } Birūpā.
MAHANADI { Mahā-nadī. { Birupā { Gengutī . } Kelo
MAHANADI { Mahā-nadī. { Birupā { Birūpā . . Birūpā . } Birūpā. } Brāhmanī
MAHANADI { Mahā-nadī. { Mahā-nadī. { Chitartala. { Chitartala } Nūn . . . . . . } Mahā-nadī.
MAHANADI { Mahā-nadī. { Mahā-nadī. { Chitartala. { Nūn . . } Nūn
MAHANADI { Mahā-nadī. { Mahā-nadī. { Mahānadī - } Mahānadī } Sukpaikā } Mahā-nadī. } Mahā-nadī.
MAHANADI { Mahā-nadī. { Mahā-nadī. { Paikā . . } Mahānadī } Mahānadī } Mahā-nadī.
MAHANADI { Kātjurī { Kātjurī . { Kātjurī. . } Kātjurī . { Kātjurī (closed by an embankment.)
MAHANADI { Kātjurī { Kātjurī . { Suruā . . } Kātjurī . { Devī . { Numerous interlacings } Devī.
MAHANADI { Kātjurī { Koyākhai (supplies Purī district, and after many bifurcations finds its way into the Chilká Lake).

Dhāmra, Brāhmanī, Mahā-nadī and Devī } BAY OF BENGAL.

The Mahānadī. The Mahānadī (the great river) has a catchment basin of 48,200 square miles, and is by far the largest of all the rivers which water the plains of Orissa. Taking its rise in the mountainous country of the Central Provinces, it emerges from the Tributary States and pours down upon the delta at Narāj, about 7 miles west of the town of Cuttack. It traverses the district from west to east and throwing off numerous branches on its way falls into the

Bay of Bengal by several channels near False Point, in 20° 18′ N., and 86° 43′ E., after a course of 529 miles.

During its progress through the hill country, it receives a vast number of streams and tributaries from the high land on either banks, but no sooner does it reach the plains than its character changes. It now forms a great delta-head, and instead of receiving confluents it shoots out a hundred distributaries. At Narāj it bifurcates, the southern branch being known as the Kātjurī, while the northern retains the name of the parent stream. The town of Cuttack is built on the spit which separates the two rivers, and opposite the town the Mahānadī proper throws off a large branch known as the Birūpā. Flowing in a north-easterly direction for about 12 miles, the Birūpā gives off the Genguti on its left bank, the two streams enclosing between them the island of Kuhunda Jaipur; the Gengutī meets the Kimiriā, an offshoot of the Brāhmanī, about 18 miles below the point of bifurcation, and rejoins its parent stream a mile lower down. After receiving the waters brought down by these streams and by the Kelo, another branch of the Brāhmanī, the Birūpā discharges itself into the main stream of the Brāhmanī a little above Indpur, and their united waters ultimately find their way into the Bay of Bengal by the Dhāmra estuary. Just below the bifurcation of the Mahānadī and the Birūpā both rivers are dammed by anicuts which control the supply of water to the head sluices of the High Level and Kendrāpāra canals.

After passing Cuttack, the Mahānadī divides into three branches, the Chitartala to the north, the Mahānadī in the centre and the Paikā to the south. The Chitartala branch leaves the parent stream about 10 miles below the Birūpā mouth, and soon bifurcates into the Chitartala and the Nūn. These streams unite after a course of about 20 miles, and under the name of the Nūn, fall into the Mahānadī estuary a few miles from the coast, and so into the Bay of Bengal. Ten miles from the point of bifurcation the Paikā and Mahānadī rejoin, but only to separate again into the Sukpaikā and Mahānadī; these branches reunite further down, and then dividing into a number of channels pass into the Bay of Bengal.

The Kātjurī, which, as already stated, is an arm of the Mahanadī, branches off at Narāj and then immediately divides into two, of which the southern branch, known as the Koyākhai or crow's pool, passes into the district of Purī; its mouth is closed by a bar, so that little water flows into it except at flood time, and it is practically a spill-channel of the Kātjurī. A short way below Cuttack the main stream throws off the Suruā, which however The Kātjuri.

rejoins it after a course of a few miles. A little lower down the Kātjurī divides again and throws off a large distributary, the Devī, down which the main body of its water passes. During its course to the sea this latter river gives off a great number of branches, the Kandal, Khandia, Dhobikhai, Purāna Devī and other minor channels, which all eventually reunite, and flow, under the name of the Devi, through the Puri district into the Bay of Bengal. The two northerly branches of the Kātjurī, the Alankā and the Kātjurī proper, have on the other hand been cut off at their head by the Devī left embankment, and the water originally carried by them has been diverted into the Devī and Tampua. The Kātjurī is said to have been originally* a comparatively small stream, and its name implies that it could at one time be crossed by a plank. During the last century, however, the volume of water passing down its channel increased considerably, and the head of the river became so much enlarged that it could not carry off the enormous volume of water poured into it by the Mahānadī. A weir and training embankment were therefore constructed at Narāj between 1860 and 1865, in order to regulate the flow and divert some of the water to the Mahānadī channels.

The Brāhmanī. The Brāhmanī, which has a catchment basin of 13,700 square miles, is formed by the junction of the South Koel and Sankh rivers in the Gāngpur State. After passing through the Bonai, Tālcher and Dhenkānāl States, it enters Cuttack district near Jenāpur, where it is crossed by an anicut. It then follows a winding easterly course, and reaches the Bay of Bengal by two mouths, the Dhāmra estuary and the Maipāra river, in 20° 47′N., and 86° 58′E., 260 miles from its source. The principal branch of the Brāhmanī is the Kimiriā, which takes off on its right bank opposite Rājendrapur village, and after meeting the Gengutī, Kelo, and Birūpā, falls again into the parent stream near Indpur under the name of the Birūpā. On its left bank, the Brāhmanī throws off the Kharsuā, which again divides into the Kharsuā and Patiyā; but the two channels reunite a little lower down and fall into the Dhāmra. As it approaches the sea, the Brāhmani mixes its waters with those of the Baitaranī, and the united stream forms a noble estuary known as the Dhāmra river.

The Baitaranī. The Baitaranī, rising among the hills in the north-west of the Keonjhar State, flows first in a south-westerly and then in an

* From the fact that clay is found at 25 to 35 feet below mean sea-level, it appears that this channel is of great age. In the case of the Koyākhai clay is met with just above mean sea-level indicating a somewhat later origin; and in the greater part of the main channel of the Mahānadī only sand has been found, from which it may be inferred that this is the main channel of the river.

easterly direction, forming successively the boundary between the Keonjhar and Mayūrbhanj States and between Keonjhar and Cuttack. It enters this district near the village of Bālipur; and after flowing in a winding easterly course across the delta, where it marks the boundary-line between Cuttack and Balasore, it joins its waters with the Brāhmanī, and passing by Chāndbāli finds its way into the sea under the name of the Dhāmra river. The principal branches thrown off from the right bank of the Baitaranī are cross streams connecting it with the Kharsuā, the chief of which is the Burha. It is navigable as far as Olokh, 15 miles from its mouth; but beyond this point it is not affected by the tide, and is fordable during the hot season. This river is the Styx of Hindu mythology, and legend relates that Rām, when marching to Ceylon to rescue his wife Sīta from the ten-headed demon Rāvana, halted on its banks on the borders of Keonjhar; in commemoration of this event large numbers of people visit the river every January.

ESTUARIES AND HARBOURS.

The rivers of Cuttack find their way into the sea by four great estuaries. On the north, the Baitaranī and Brāhmanī debouch into the Bay of Bengal at Palmyras Point, by the two mouths known as the Dhāmra and Maipāra; and after numerous ramifications the Mahānadī, or rather that portion of it which remains in Cuttack district, forms two great estuaries, one generally known as the Devī, in the south-eastern corner of the district, while the other, bearing the name of the parent river, empties itself into the sea at False Point, about halfway down the coast. In spite, however, of the existence of these estuaries and of the extent of its sea face, the district does not contain a single harbour capable of sheltering ships of any great size. An eternal war goes on between the rivers and the sea on the monsoon-beaten coast, the former struggling to find vent for their columns of water and silt, the latter repelling them with its sand-laden currents. These forces counteract each other, and the sea deposits a bar outside the river mouth, while the river pushes out its delta to right and left inside. All the estuaries therefore have a bar of sand across the mouth, which prevents the entrance of vessels of any large burden except at high tide; most of the trade passes through Chāndbāli in the Balasore district at the mouth of the Baitaranī; and False Point, where ships can ride in an exposed roadway, constitutes the only anchorage on the coast. The following is a brief description of each of the estuaries.

The Devī estuary.

The Devī, with its channel, the Jotdār, forms the last part of the great network of rivers into which the Kātjurī branch of the Mahānadī bifurcates. According to a characteristic of the Cuttack streams, most of the members of this network reunite as they approach the ocean; and the result is a broad estuary, known as the Devī, which enters the sea a short distance to the south of the boundary between Cuttack and Purī. This estuary is navigable up to Māchgaon by small sloops, which use this channel to obtain cargoes of oil-seed and rice. It is one of the best tidal channels in Orissa, but owing to the bar of sand at its mouth vessels of large size cannot enter it except at high tide.

The Mahānadī estuary.

The northern branches of the Mahānadī also join as they approach the sea, and eventually enter the Bay of Bengal under the name of the parent stream. The estuary has several mouths, but the principal one is that which debouches through the shoals to the south of the False Point lighthouse. For many miles up the river, there is abundance of depth for ships of large burden; but unfortunately, as in the case of Devī, and indeed of all other Orissa harbours, a bar stretches across the mouth, which, in addition to the perils of shoal water, adds the dangers incident to constant changes in the channels and the sandbanks. The False Point harbour, which lies a little north of the Mahānadī estuary, is a comparatively exposed anchorage, and loading and unloading cannot be carried out in rough weather. Two separate channels lead inland from the anchorage, on the north the Jambu river, and on the south the Bākuḍ creek, a short branch of the Mahānadī. A more detailed description of the harbour will be found in Chapter XV.

The Brāhmanī estuaries.

The river system of the Cuttack district on the north of the Mahānadī consists of the network of channels formed by the Brāhmanī and Baitaranī, which, after infinite windings, find their way into the sea by two great outlets at Point Palmyras. The southern of these is the Maipāra river, with its tidal creek, the Bānsgarh, which runs southward almost parallel to the coast till it joins the sea about 6 miles north of False Point harbour. The mouth of the Maipāra presents the usual obstacles of bars and high surf, and from its position on the south of the Palmyras promontory, it is inadequately protected from the monsoon. Between the months of November and March this last objection does not apply, and native craft from the Madras coast frequent it during the cold weather for the purchase of rice.

The Dhāmra, the northern exit of the united streams of the Brāhmanī and Baitaranī, forms the boundary-line between the districts of Cuttack and Balasore, but is within the jurisdiction of the latter district. The Dhāmra, though navigable, is rendered dangerous by a bar across its mouth; but the entrance has greatly improved of late years, and at flood tide vessels drawing as much as 18 feet can pass in with safety.

GEOLOGY.

The greater portion of the district consists of alluvium. Much of this is the recent deltaic deposit of the Mahānadī and Brāhmanī rivers, and occupies a tract of country extending some 30 miles from the sea, which is perfectly flat with the exception of a belt of low sand hills along the coast. In the north-western part some undulating ground consists of an older alluvium containing *kankar* and pisolitic ferruginous nodules, with no defined boundary line to separate it from the inland laterite on the one hand and from the alluvium of the delta on the other. This laterite is evidently of detrital origin, and consists of small pisolitic nodules of hæmatitic iron, and coarse quartz sand. The extent to which these cohere varies greatly. The variety used as building stone possesses the property of being quite soft and easily cut when first dug, but hardening on exposure. It occurs around all the gneissose hills near and to the west of the road from Calcutta to Cuttack, forming broad terrace-like flats stretching from hill to hill, except where it is concealed from view by the ever-lying alluvium.

Starting from the district boundary north of the town of Cuttack and extending all along the border between Cuttack and the Tributary States of Orissa, are hills belonging to the Archæan crystalline group, which is so strongly developed throughout Southern and Eastern India. Between the Brāhmanī and Mahānadī rivers these hills are more or less isolated and all composed of gneiss, in places compact and granitoid, in others partially disintegrated and marked with numerous red blotches, the remains of decomposed garnets; this soft decomposed gneiss is sometimes quarried and used for building. North of the Brāhmanī river our information is less definite, but it is probable that the hill ranges consist of rocks belonging to the same crystalline group. The only other rock formation is found in a large area, to the west and south-west of the town of Cuttack, occupied by grits, sandstones and conglomerates with white or pink clay beds. These are known as the Cuttack or Athgarh sandstones. They are bounded by laterite towards the Koyākhai river and extend beyond the district boundaries into Athgarh and Purī. They are unfossiliferous, but their lithological resemblance to the Pānchets of Upper Gondwāna age has led to their being classed as such, although they are disconnected with the Gondwāna rocks exposed in the Tālcher coal-field.

For fuller details of the geology* of the district, the following papers may be consulted: "The Geological Structure and Relations of the Tālcher Coal-field" by W. T. and H. F. Blanford and W. Theobald, jun.; "The Geology of Orissa, etc.," and "The Laterite of Orissa" by W. T. Blanford, in the "Memoirs of the Geological Survey of India," Vol. I; also "A Sketch of the Geology of Orissa" by W. T. Blanford in the Records of the Geological Survey of India, Vol. V, Part II.

BOTANY. The banks of the sluggish rivers and creeks which wind through the swampy low-lying country near the sea exhibit the vegetation of a mangrove forest, the principal species being *Rhizophora*, *Ceriops*, *Kandelia*, *Aegiceras*, *Hibiscus tiliaceus*, *Excoecaria Agallocha*, and the like. Where sand dunes intervene between the sea and the cultivated land behind, a littoral vegetation uncommon in Bengal is met with, which includes *Spinifex*, *Hydrophylax*, *Geniosporum prostratum* and similar species. These sand hills stretching between the fertile rice plains and the sea constitute the only really distinctive feature of Orissa from a botanical point of view, and present not a few of the littoral species characteristic of the Madras sea-coast. The cultivated land which occupies the sub-montane tract to the west has the usual rice-field weeds such as *Ammannia*, *Ilysanthes*, *Dopatrium*, *Utricularia*, *Sphenoclea*, *Hygrophila*, while ponds and ditches are filled with floating water weeds like *Pistia*, *Trapa*, *Nymphæa*, *Limnanthemum*, or submerged water plants, such as *Hydrilla*, *Ceratophyllum*, *Vallisneria* and *Ottelia*. Near human habitations shrubberies containing *Trema*, *Glycosmis*, *Polyalthia suberosa*, *Triphasia*, *Adhatoda Vasica*, *Solanum torvum* and *verbascifolium*, *Clerodendron infortunatum* and such like semi-spontaneous shrubs are common. This undergrowth is loaded with a tangled mass of climbing *Naravelia*, various *Menispermaceæ*, many *Apocynaceæ*, several species of *Vitis*, a number of *Cucurbitaceæ*, *Basella*, and several *Convolvulaceæ*. The arborescent portion of these village-shrubberies includes the red cotton tree (*Bombax malabaricum*), *Odina Wodier*, *Tamarindus indica*, *Moringa pterygosperma*, the *pīpal* (*Ficus religiosa*), the banyan (*Ficus bengalensis*), the palmyra (*Borassus flabellifer*) and the date palm (*Phœnix sylvestris*).

In the north-western portion of the district other species of a more truly forest character are found, among them being *Ailanthus excelsa*, *Pterocarpus Marsupium*, *Pterospermum Heyneanum*, *Dalbergia paniculata* and *lanceolaria*, *Melia composita*, *Adina cordifolia*,

---

* The account of the geology of the district has been contributed by Mr. G. E. Pilgrim, Deputy Superintendent, Geological Survey of India.

Schleichera trijuga, and the like. The bamboo of the district is usually *Bambusa arundinacea.* Open glades are filled with grasses, sometimes of a reedy character; sedges are abundant; and ferns are fairly plentiful.

There are no forests in the district, but wide stretches of jungle extend along the sea coast and in the hilly border to the west. The woodland strip on the sea-board furnishes supplies of fire-wood and materials for cottage-building and agricultural implements; while large quantities of bamboos and fuel are brought down from the hilly tract adjoining the plains. *Sāl* trees are found in *Kilā* Sukindā, the proprietor of which carries on a trade in timber; and other products extracted from these western jungles are resin, wax, honey and nux vomica. Forests.

Wild animals are still plentiful in the district, in spite of the extension of cultivation and the construction of the canals, and some of the larger species of carnivora abound in the hilly region to the west and in the low-lying country near the sea-board, where the dense jungle has not yet given way to the plough. Tigers are found in nearly all the large tracts of jungle, especially near the coast and in the northern part of the district, but they are occasionally found killing close to the town of Cuttack itself, as the Mahānadī, lying between the station and the hills, is an effective barrier to the advance of civilization. They are responsible for comparatively little loss of cattle or human life near the coast, where the population is sparse and there is plenty of game, such as deer and pig, to supply them with food. In other parts, however, they annually cause considerable destruction, and in th four years 1901-04 they killed 67 persons and over 3,000 cattle; in the same period 63 tigers were shot and their dead bodies brought in for rewards. The jungle which they frequent is generally dense, and beats are, as a rule, unsuccessful, while, owing to the large tract of country over which they range, it is difficult to get information in time, and shooting from *machāns* is, therefore, rarely attempted. Leopards are found all over the district; bear, spotted deer, hog-deer, *sāmbar* (*Cervus unicolor*) and mouse-deer are common in the interior; while black buck have their habitat on the sandy tracts along the coast. Bison are now rare, but stray wild elephants occasionally wander across the borders from the Garjāts. Boar are found in large numbers along the coast near Jambu, and porcupine, jackals and foxes are numerous. Wild dogs are occasionally met with, but wolves have now disappeared. FAUNA.

Indigenous quail, hare and black partridge are found in the scrub jungle, and migratory quail also occasionally visit the district. Pea fowl and jungle fowl are common, and nearly every variety of wild duck is found. Snipe are numerous but scattered, and large bags are rare. They are seldom found in the standing paddy as in Bengal and Bihār, but prefer jungle grass in the earlier part of the season, while in January and the following months they are found in wisps in the paddy stubble. Numbers of native *shikāris* shoot, snare and net for the local markets; they destroy wholesale and are rapidly diminishing the number of game of all kinds.

Crocodiles and *ghariyāls*, or fish-eating alligators, abound in the tidal rivers and creeks, and grow to a very large size; the snub-nosed or man-eating crocodile annually levies a heavy toll on cattle and human life, and in the two years 1903-04 carried off 68 persons and 184 cattle.

A large variety of fish are found in the rivers, and the Oriyā spends all his spare time in fishing, often standing up to his neck in water for the greater part of the day. Fish, either fresh or preserved by drying in the sun, are a favourite dish; dried fish, known as *sukuā*, being particularly popular with boatmen, carters, etc. *Hilsa* are caught in season, and the prawns of Cuttack are famous for their size and delicacy. Oysters are good and plentiful on the coast to the extreme north-east.

CLIMATE. Temperature. The district is directly on the track of the cyclonic storms which frequently cross Orissa during the monsoon season, and the extremes of climate are more marked than in most other parts of Bengal. In April and May the average maximum is 102°, and at Cuttack temperature has been as high as 118°; while the mean temperature falls from 88° in the hot-weather months to 83° in the monsoon season and to 69° in February. It is one of the hottest districts in the Province, and the account of William Bruton, one of the small band of Englishmen who first visited it in 1633, shews how intensely they felt the heat. On the 28th April, he writes: "At the hours of between eleven and twelve of the clock, it was so excessively hot that we could not travel; and the wind blew with such a sultry scalding heat as if it had come forth of an oven or furnace; such a suffocating fume did I never feel before or since." The cold weather commences in the beginning of November, the temperature begins to cool, and the mornings and evenings are chilly, though the air has not the same bracing invigorating effect as in Northern India. With March the heat approaches, and by April the hot weather has fairly set in; during these two months, and frequently in May, there are occasionally showers of rain accompanied by strong north-westers. Throughout these three months the heat is excessive in the day-time, but no sooner has the sun set than a strong sea-breeze prevails, and a punkah can almost be dispensed with, when it is in full force. Formerly the European residents used to repair to the coast during the hot season of the year, and an early account describes Purī as being to the residents of Cuttack what Brighton, Margate and other sea-bathing quarters are to the inhabitants of London. The rainy season begins in June or early in July, and the rains last till the end of September or the month of October, when an unpleasant time of moist heat marks their cessation.

Owing to the dry westerly winds which occasionally sweep across the district in the hot season and to the well-marked south-west monsoon conditions which occur later in the year, humidity undergoes considerable variation, ranging on an average from 72 per cent. of saturation in April and May to 83 per cent. in August. Humidity.

The cyclones which occur in the rains proper (*i.e.*, in June, July, August, and September) are generally small in extent, the barometric depression at the centre seldom exceeding half an inch, and the air motion, though violent, is rarely of hurricane force. The district is not liable to suffer from the devastating cyclones which occasionally occur in the months which precede and follow the full establishment of the south-west monsoon, *i.e.*, during April and May, October and November, as the tendency is for such storms to move into the north of the Bay and recurve towards the Arakan or Bengal coast. But in recent years several severe cyclones have been experienced, the most notable being that of the 23rd September 1885, which caused great havoc in the neighbourhood of False Point. Cyclones.

The normal annual rainfall is 60·43 inches, of which 4·6 inches fall in May, 9·9 in June, 11·7 in July, 12·3 in August, 10·3 in September and 5·8 in October. From June to September the monthly rainfall varies from 10 to 13 inches on an average, with considerable fluctuations from year to year, according as the cyclonic storms are more or less numerous and move in the usual course westward over Orissa. In October the rainfall depends on causes similar to those mentioned below for May and is similar in amount, averaging 4 to 5 inches. Between November and April rainfall is light and is usually caused by local thunderstorms. Cyclonic storms occasionally occur in the north of the Bay of Bengal in May, and with these storms weather of the south-west monsoon type prevails. An extreme case of such rainfall occurred in 1893 when more than 20 inches, or one-third of the normal annual amount, fell in Rainfall.

this month owing to a prolonged cyclonic disturbance in the north-west angle of the Bay.

Statistics of the rainfall for the various recording stations are given below for the cold weather (November to February), the hot weather (March to May) and the rainy season (June to October). The figures shewn are the averages recorded from the earliest year in which rainfall was systematically registered up to the end of 1905 :—

STATIONS.	Years recorded.	November to February.	March to May.	June to October.	Annual average.
CUTTACK ...	40—43	2·52	6·23	51·60	60·35
BANKI ...	18	2·57	4·94	47·24	54·75
DHARMSHALA ...	15—16	2·37	8·13	49·34	59·84
FALSE POINT ...	33—34	4·60	7·21	57·62	69·43
JAGATSINGHPUR	28—30	3·29	5·54	49·04	57·87
JAJPUR ...	28—31	2·86	8·02	50·63	61 51
KENDRAPARA ...	30—31	3·33	6·78	51·03	61·14
SALIPUR ...	14—16	2·42	6·54	49·46	50·42

CHAPTER II.

HISTORY.

ORISSA* practically emerges into the light of history in the rock edicts of Asoka (B.C. 260). The references to its early inhabitants which can be traced in traditional or legendary chronicles are extremely few, and the remains of an anterior date which have been discovered are still fewer. Scanty, however, as are the materials from which any idea of its original inhabitants can be obtained, they are sufficient to justify the inference that in prehistoric times the hills were peopled by savage tribes differing from those which occupied the low lands, and that the intervening plains were in the possession of races somewhat more civilized. The only remains of the stone age hitherto found are some roughly chipped quartzite axes, discovered in Dhenkānāl, Angul, Tālcher and Sambalpur. Similar axes have been found in considerable numbers in the Madras Presidency, and to a smaller extent in the Central Provinces; and this fact may point to some connection between the tribes living in these localities and those dwelling in Orissa during the stone age. PREHISTORIC PEOPLES.

It is probable that several of the tribes which still inhabit the hilly country to the west were originally natives of Orissa; but here, as in other parts of India, the absence of reliable data makes it difficult to separate the later immigrants from the early settlers. According to the traditions current among those tribes, the Khonds of the south, the Gonds of the west, and the Hos, Bhumijes and Santāls of the north would appear to have migrated into Orissa in historic times. The Bhuiyās of Keonjhar allege that they are autocthonous, but the Juāngs deny this; and other Bhuiyās have been found in Rānchī and other parts of Chotā Nāgpur who claim to have been originally settled there.

---

* Properly speaking, Orissa (Sans. *Utkala*, vernacular *Odīsā*) means the tract in which the speakers of Oriyā form the dominant people. During the period of British rule the name has been applied to the tract extending from the Chilka Lake to the river Subarnarekhā, and comprising the districts of Purī, Cuttack, Balasore, and Angul, besides 17 Tributary States. Unless the context shews otherwise, this will be the meaning of Orissa in this chapter.

The Savars

The Savars, who in Cuttack hold a degraded position as hewers of wood, have better claims to be regarded as an indigenous tribe of Orissa. In the Aitareya Brāhmana and Sānkhāyana Srauta Sūtra, they are mentioned as one of the degraded races (the Andhras, Pundras, Savars, Pulindas and Mutibas) descended from the fifty sons of the sage Visvāmitra, who were cursed by their father for being dissatisfied with his acknowledgement of Sunahsepa as his first-born. They are several times alluded to in the Bhagavatī, the oldest sacred literature of the Jains, where their language is referred to as one of the *Mlechchhabhāshās* or barbarous tongues; and they have been identified with the Suari of Pliny and the Sabarai of Ptolemy. In the Mahābhārata, the Savars are placed in the Dakshināpatha, *i.e.*, the region to the south, and in the Brihat Samhitā in the south-east of India, and this is confirmed by Pliny and Ptolemy. Their geographical distribution has not been much changed, and they are still found in Midnapore, Singhbhūm and Orissa.

The Juāngs.

The Juāngs of the Tributary States, who are one of the most primitive races of India, would seem to be another of the early tribes of Orissa. Till they were clothed by order of the Government, the only covering of the females consisted of a few strings of beads round the waist, with a bunch of leaves before and behind—a practice which has given them the name of Patuas or Patrasaras (leaf-wearers) in Orissa; they had no knowledge of the metals till the 19th century, when foreigners came among them; and no word existed in their own language for iron or any other kind of metal. But their country abounds in flint weapons, and it has been suggested that they are the direct descendants of the ancient stone-cutters. They may be the Parnasavaras of the Mārkandeya Purāna, and the Drillophylletes of Ptolemy.

The Pāns.

The Pāns, who are found scattered throughout Orissa, Singhbhūm, Rānchī, and the adjoining tracts in the Central Provinces and Madras, should also probably be regarded as one of the prehistoric peoples of Orissa. Everywhere they rank among the lowest classes; they are employed in servile occupations even by such tribes as the Khonds and Bhuiyās; and in the days of human sacrifices, the Khonds selected a Pān boy as the best sacrifice which could be offered to mother earth. All these facts seem to indicate that they were the original occupants of the soil, who were dispossessed and reduced to slavery by other tribes. They are possibly the Parnnakas of the Vājasaneyi-Samhitā and the Taittirīya Brāhmana, and the Nagna-parnas of the Brihat Samhitā mentioned as living in south-eastern India.

The sea-coast and the low lands behind it were presumably occupied by tribes who followed the occupations of fishermen and boatmen. The Kewats, including the cognate castes, the Gokhās and the Mallāhs, have been traced to very early times as the Kevata in Asoka's Pillar Edict No. V, and, in the Sanskrit form of the name, as the Kaibarttas mentioned in the Vājasaneyi-Samhitā and the Taittirīya Brāhmana. The Kewats.

The intervening plains and uplands appear to have been held by tribes on a somewhat higher level of civilization. From the scanty references made to them in later literature, it would seem that some of these tribes were known as Odras and Utkalas. The Odras and the Utkalas appear as different tribes: the former are now and then joined with the Paundras in the Mahābhārata, the Manu Samhitā and the Brihat Samhitā (*Paundra-Odra*), and with the Marundas in the Bhagavatī (*Marùnd-Odra*), while the Utkalas are connected with Mekalas in the Mahābhārata, the Rāmāyana and the Brihat Samhitā. The Paundras occupied the land now known as the Rādha; while Mekala was the tract round about the Amarakantaka hills, the river Son being specially described in the Harivansa as Mekala-prabhava. If this connection has any significance, it means that the Odras occupied the eastern and the Utkalas the western side of the country. A tradition in the Harivansa speaks of Utkala and Gayā as being relatives; and in the Mahāvagga, Tapussa and Bhallika, the first lay-disciples of Buddha, are said to have come to Bodh-Gayā from Utkala. In course of time, they spread southwards, until the Utkalas were absorbed in the larger tribe of Odras, though they gave their name to the land in Sanskrit works at least before the 5th century A.D. Gradually they spread further south to Kalinga, till that land became divided between two main speeches, the Oriyā and the Telugu. The Utkalas and Odras.

It seems probable that before the 3rd century B.C., several of the Indo-Aryan castes, such as the Brāhmans, Kshattriyas, Karans and others, had migrated to Orissa, which then formed part of Kalinga. In the Bandhāyana Dharma Sūtra it is laid down that he who has visited Kalinga must offer a sacrifice in penance; and in support of this an older verse is quoted. In the Mahābhārata pilgrims are asked to avoid Kalinga; it also says that the Kshattriyas in Kalinga had become outcastes; and a similar statement is made in the Manu-Samhitā regarding the Kshattriyas who lived among the Odras. These references appear to point to the migration of several Indo-Aryan castes, and among them there must have been Brāhmans, though they were not recognized as such in the Madhyadesa. The Māstāns Early Indo-Aryan settlers.

and the Sāruās are probably the descendants of these early immigrants; they call themselves Brāhmans, and wear the sacred thread, though they neglect the nine *sanskāras* or ceremonies incumbent on Brāhmans, and have taken to forbidden occupations, such as cultivating with their own hands, selling vegetables, etc.

EARLY HISTORIC PERIOD.

Kalinga.

As Orissa formed part of Kalinga before the conquest of Asoka, its early history is merged in the history of that country. Kalinga extended, according to the Mahābhārata, southwards from the *Gangā-Sāgara-sangama* or the junction of the Ganges with the sea, the river Baitaranī in Orissa being specially mentioned as in Kalinga; while, according to Pliny, it stretched as far south as the promontory of Calingon, which is identified by Cunningham with the promontory of Coringa at the mouth of the Godāvari. It was an extensive, populous and fairly civilized kingdom. Some idea of the vast number of its population may be gathered from the Rock Edict XIII, which begins with saying that when Asoka conquered Kalinga, 150,000 persons were carried away captive, 100,000 were slain, and many times that number perished. The evidence of the high standard of civilization and prosperity attained in Kalinga is equally striking; elephants were specially bred for the royal forces, of which they formed a prominent part; diamonds of a special kind were quarried and exported; there was an entirely separate measure for medicines; cloth was manufactured and exported in such quantities that Kalinga became the word for cloth in old Tamil; and frequent sea voyages were made to countries outside India, on acccount of which the Indians came to be called Klings in the Malay Peninsula.

Asoka's reign.

As the result of the bloody war mentioned above, Orissa and Kalinga were incorporated in the empire of Asoka in the ninth year of his reign, *i.e.*, in 262 or 261 B.C. The horrors which accompanied this war made a deep impression on the heart of the victorious monarch, who recorded on the rocks in imperishable words the sufferings of the vanquished, the remorse of the victor, and his conviction that the only true conquest is that effected by *dharma*, or the law of piety, and not by force of arms. The rock edicts which he had inscribed in Orissa consist of General Edicts I—X and XIV, and Special Edicts I and II, incised upon the rocks at Dhauli in the district of Purī and at Jaugada in Ganjām. From the special edicts it would appear that the Empire was divided, for administrative purposes, into several great divisions, a prince being placed in charge of each, with the exception of the head-quarters division, containing the capital at Pātaliputra, which was under the direct supervision of the Emperor; and it was considered necessary to place the extensive and newly conquered territories of Kalinga under a Viceroy stationed at Tosāli (Sep. Edict II). Tosāli, to the officers in charge of which the Dhauli separate edicts were addressed, and Samāpā, to the officers in charge of which the Jaugada separate edicts were addressed, must have been towns not far from the rocks. Hence Tosāli was presumably some place close to the modern Bhubaneswar, which is not far off from Dhauli, has old remains in the neighbouring caves, and from its upland position commanding the bifurcations of the rivers was well fitted for the site of a capital town. Samāpā would similarly seem to have been a town on the river Rishikulyā not far from the modern town of Ganjām.

The Maurya rule in Kalinga.

According to the Purānas, the Mauryan Empire lasted till about 180—170 B.C., and Orissa was therefore under the sway of its kings for 80 to 90 years. During this time it must have come into closer relations with Northern India. Its inaccessibility was to some extent removed by roads lined with banyan and mango groves, with wells and rest-houses, and by the arrangements made for the greater safety of Government messengers and travellers. These measures naturally facilitated an influx not only of officials but also of traders and pilgrims, some of whom eventually settled in the land. Hence in the Mahābhārata, one finds later verses declaring that there were good men in Kalinga, and that *tīrthas* existed in that country, thus withdrawing the ban laid on travelling there. With the Jainas Kalinga ranked still higher as an *Ariya* country, and naturally so, for one finds traces of their very early residence in the land. The sandstone hills of Udayagiri and Khandgiri, 5 miles north-west of Bhubaneswar, are honeycombed with caves, all of which appear to have been made by the Jainas. Of these caves, the Sarpa, Bāgha and Jambesvara caves at Udayagiri, and the two Tātuā caves at Khandgiri have short inscriptions in the Brāhmi character of the Mauryan age (3rd century B.C.).

Reign of Khāravela.

Vrihadrath was, according to the Purānas, the last of the Mauryas. He was dethroned by his general Pushyamitra, who founded the Sunga dynasty about B.C. 180-170; and his overthrow having brought about the disruption of the empire, Kalinga again became independent. This is evident from the interesting but mutilated inscription on the top of the Hāthi Gumpha or Elephant Cave at Udayagiri, which is dated in the 165th year of the Maurya rule, *i.e.*, 153 B.C., and purports to narrate the career of Khāravela, king of Kalinga, up to the thirteenth year of his reign. From this inscription we learn that

Khāravela, *alias* Bhikurāja, the son of Vadharāja and grandson of Khemarāja, of the Cheta royal family, regained the city and fort of Kalinga in the first year of his reign. In the second year he sent a large army to the west to protect Sātakarni, and with the help of the Kusāmba Kshattriyas captured the town of Māsika. In the eighth year he made an expedition against Rājāgriha, whose king fled to Mathurā; next year he harassed the kings of the north, and in the twelfth year he again invaded Magadha and made its king his vassal. Besides this account of his military power, the inscription records the pious deeds of the king, his repair of an alms-house built by Nandarāja, his gifts to Brāhmans and Arhats, the musical entertainments he provided for the people, the assembly of sages he called together, the construction of an aqueduct, and the additions he made to a *chaitya* set up by his predecessors.

This inscription shews that Khāravela made Kalinga a powerful empire. One of his first acts was to assist Sātakarni, king of the Andhra country, *i.e.*, the tract between the Godāvari and the Krishnā, in fighting against his suzerain, the Sunga Emperor, Pushyamitra, in 164 B.C. The invasion of Magadha and its old capital Rājāgriha later in his reign indicates that the Kalinga king had become not only independent but aggressive. In this war a successful expedition into the heart of the empire led him to the capital, Pātaliputra, on the banks of the Ganges, and compelled the Emperor to sue for peace and acknowledge his independence. The invocation and the mention of Arhats in the inscription afford good grounds for the belief that the king and his family had a leaning towards Jainism, and this is borne out by the inscription in the Svargapurī cave which says that the cave was made for the Arhats by Khāravela's queen. His successors appear also to have been adherents of Jainism, to judge from the inscriptions they have left in these Jaina caves, as the first inscription in the Manchapurī cave describes it as being made by Vākdepa, who has the same three titles as Khāravela and was probably his successor, and the second inscription describes the cell as the cave of Kumāra Vadukha, who may be the son of Vākdepa.

The Andhra rule.

It is not known how long this dynasty lasted or by whom its kings were succeeded. But in the second century A. D., Kalinga appears to have been overshadowed and probably absorbed by the Andhras. Ptolemy in his map shows the whole sea-coast up to the Ganges as included in Maisolia or the Andhra country. The introduction of Buddhism, as handed down by tradition, may be ascribed to the same cause; and it is noticeable that the Tibetan chronicles have preserved a tradition that Nāgārjuna converted Manja, king of Otisha, to Buddhism with one thousand of his subjects. Nāgārjuna is believed to have flourished about 200 A.D., in the court of the Sātavāhana, *i.e.*, the Andhra royal family; and the conversion of the people to Buddhism would naturally have been facilitated, if Orissa was subject to that powerful dynasty.

The Imperial Guptas.

After this there is another gap until one comes to the Imperial Gupta dynasty. In Samudra Gupta's Allahābād pillar inscription, it is said that he captured and then liberated various kings of the Dakshināpatha, including those of Kosala, Kerala, Pishtapura, Kānchī, Vengī and others. If this be accepted as a record of fact, Orissa must have been conquered, but it is doubtful whether the expressions are not hyporbolical; and this supposition is borne out by the fact that not a single monument or further mention has yet been found of the Gupta conquest of the Deccan.

THE MEDIÆVAL PERIOD.

From an inscription of the Sailodbhava Madhavarājā II, dated 619—20 A.D., we learn that Kongeda, a country which may be identified with the strip of land south of Orissa, had come under the sway of Sasānka, the powerful king of Gauda, by the beginning of the 7th century, and Orissa must also have acknowledged his sovereignty. Shortly afterwards both the countries were conquered by Silāditya Harshavardhana of Kanauj. The precise date of his conquest is not known, but from the life of Hiuen Tsiang it would appear that he had conquered Orissa before the death of Purnavarman, king of Magadha, as he offered the Buddhist monk, Jayasena, the rental of 80 villages in Orissa in order to induce him to come to his Court, and that he had conquered Kongeda before he met the priests of Orissa who decried the Nālanda doctrines as the "sky-flower" system, It was during Silāditya's reign that the Chinese traveller, Hiuen Tsiang, visited Orissa (639 A.D.), and we have a short but graphic account of the country in his records. The country, he says, was about 7,000 *li* (a *li* is $\frac{1}{5}$th to $\frac{1}{6}$th of a mile) in circuit, the climate was hot, the soil was fertile, and produced abundance of grain and fruit. The people were uncivilized, tall of stature and of a yellowish-black complexion. They loved learning and applied themselves to it without intermission. Most of them believed in the law of Buddha. There were some hundred monasteries with 10,000 priests, all studying the Mahāyāna or the Great Vehicle, and 50 Deva temples were frequented by sectaries of all sorts. The capital, which has been identified with Jājpur in this district, lay 700 *li* south-west of Tamrālipti (Tamlūk); on the south-west frontier was a

Silāditya.

miraculous monastery, called Pushpagiri, situated on a great hill; and on the south-east frontier, on the borders of the ocean, lay a great walled port named Charitra.

The Kesarī dynasty.

On the death of Silāditya, his empire was dismembered, and, according to the Mādalā Pānji or palm-leaf chronicles of the temple of Jagannāth, Orissa was under the Kesarī or Lion dynasty from the 7th to the 12th century A.D. The very existence of this dynasty is denied by several scholars, but in the Bhakti-bhāgavata Mahākāvyam, a Sanskrit poem of 1409-10 A.D., which gives a very brief history of Orissa, it is distinctly stated that the Kesarī kings preceded the Gangas and that Udyota Kesarī was one of them; two inscriptions of the time of Udyota Kesarī have been discovered, one in the Nabamuni cave on the Khandgiri hill and the other in the Brahmeswar temple at Bhubaneswar. M. Sylvain Levi, moreover, states that in the Japanese edition of the Chinese Tripitaka is a translation of a part of the Buddhist Buddhāvatamsaka Sūtra, made by a monk in 796-98 A.D. on a copy of the Sūtra which was sent as a present to the Emperor of China by the king of U-tcha (Odra), and that this king is named in the letter of presentation as Srī Māhesvar or Parama Māheshvar Mahārājā, doing-pure lion-king (Subhakara Kesarī). Another Kesarī king of Orissa, Karna Kesarī, is mentioned in the commentary of the historical poem Rāmapāla-Charitam as having been defeated by Jayasingh, king of Dandabhukti (Bihār): both the poem and the commentary are believed to be by the same author, probably a contemporary of the hero of the poem, Rāmapāla, king of Magadha, who flourished in the latter half of the 11th century.

Mediæval civilization.

According to the Brahmeswar inscription, there were five Kesarī kings, the first being Janmejaya and the last Udyota Kesarī, who was king of Kalinga and defeated the Singhalas, Chodas and Gaudas. The palm-leaf chronicles attribute most of the great temples at Bhubaneswar to this dynasty, and this, if true, must place it among the most important dynasties of India. From inscriptions and other sources we learn that the fine temple of Brahmeswar was built under the orders of Queen Kolāvatī, that the smaller but elaborately-carved temple of Someswar was constructed in the time of Udyota Kesarī, and that the large temple of Ananta-Bāsudeva was erected by Bhavadeva Bhatta, probably in the 11th century. The Buddhistic remains at Naltigiri, Udayagiri and Ratnāgiri in this district are ascribed by local tradition to Vasukalpa Kesarī, the husband of Kolāvatī, who may perhaps be identified with the father of Udyota Kesarī. The number and magnificence of the remains at Bhubaneswar and elsewhere make it clear that the kings who erected these great works must have held vast and populous dominions and must have been able to command ample resources. The art of architecture and sculpture must have been well developed to enable such huge and lavishly carved structures to be designed and constructed; the artisans must have received a good training both mechanically and artistically before they could have moved and laid in place (without mortar) such gigantic stone-blocks, or could have produced the vigorous and often exquisitely carved figures, foliage and arabesque patterns, which lend a charm to the carvings adorning these shrines.

These stately temples show the hold which Hinduism had obtained in Orissa by this time, but a few centuries earlier Hiuen Tsiang found Buddhism flourishing side by side with Hinduism, and his account is corroborated by the present of Buddhist scriptures to the Chinese Emperor in the 8th century. The Mahāyāna type of Buddhism, which the Chinese pilgrim found in Orissa, seems to have been supplanted gradually by the Tāntrik forms of the Magadha school, of which traces may be found in the Bodhisatwas, Bhairavas, Tārās and Vajrayoginīs which are still to be seen in the Assia hills of this district. The caves at Khandgiri and Udayagiri just across the border of this district also bear signs of a Jaina revival, probably a reflex of the Jaina influence in the Western Deccan which began during the period of the rule of the Chalukyas and predominated in the time of the Rāshtrakūtas (748-973 A.D.). Magnificent as are these monuments, not a single literary work of this period has as yet come to light. Buddhist philosophy, however, was not neglected, as otherwise the monks would not have dared to challenge and decry the doctrines taught in the great Nālanda monastery. So far too as can be gathered from inscriptions, the study of Sanskrit was kept up; and in the Sarasvatīkanthābharana, a rhetorical work of the 11th century, a special class of alliterations is distinguished by the name Odrī.

The Eastern Ganga dynasty.

In the beginning of the 11th century, the Cholas, who had established a great empire in the Deccan, began to extend their power over Orissa, and we learn from the evidence of inscriptions that Rājarāja Rāja-Kesarīvarman conquered Kalinga between 1008-1010 A.D., and that his son Rājendra-Choladeva in 1021-1023 subdued "Odda-vishaya, whose copious waters are difficult to approach." These conquests do not appear, however, to have left any permanent mark on the country, being merely brief but successful expeditions, like the defeat of Karna Kesarī mentioned above. The rising power of the kings of Kalinga, the Eastern

Gangas of Kalinganagara (the modern Mukhalingam in the Ganjām district), was a much more serious menace to the independence of Orissa. An inscription of 1075-76 A.D. states that the king of the Odda country was one of the kings defeated by a general of Rājarāja I; Chodaganga, the son of this monarch, reconquered Orissa, and on the death of the Odra king it passed finally into his hands. A copper-plate grant of Chodaganga dated 1118-19 describes him as the sovereign over the whole of Utkala; it seems certain therefore that the conquest took place before that date, and it is probable that it occurred soon after his coronation in 1078 A.D.

The rule of the Eastern Ganga kings lasted till 1434-35, the dynasty including altogether 15 kings. Of these by far the most powerful was Chodaganga or Gangesvar, who conquered Vengī on the south and Mandāra on the north, and by these conquests extended his dominions from the Ganges to the Godāvari. The famous temple of Jagannāth was built by him, and the shrine of Gangesvar in Jājpur, which was apparently constructed under his orders, was named after him. On his death after an extraordinarily long reign of 72 years, his sons succeeded one after the other, but their reigns are almost barren of interest; and we only know that, according to a Bengal inscription, Rāghava (1156-70) was defeated by the Sena king Vijaya Senā, and that the great temple of Megheswar at Bhubaneswar was erected by the brother-in-law of the next monarch between 1193 and 1198.

Muhammadan raids.

The only notable events in the reigns of the succeeding sovereigns are their struggles with the Musalmāns of Bengal, and later on with the Bāhmani and other Sultāns. In 1205 came the first Muhammadan incursion, when Muhammad-i-Shīrān, an officer of Bakhtiyār Khiljī, burst down upon the country, and this incursion was followed by many others. In an inscription at the Jagannāth temple at Chāteswar in this district, the founder, Vishnu, a Brāhman minister of Ananga Bhīma Deva (1211-1238), claimed to have fought with Yāvanas, by which he probably means Ghiās-ud-dīn Iwaz, the fourth Bengal Sultān, and with the lord of Tummāna in the Chedi country. The Tabakat-i-Nāsiri records in 1244 first a raid made by the Orissan army, and then a counter-raid of the Bengal king Tughril-i-Tughān Khān, which ended with his defeat by the local levies; in 1245 the Oriyās retaliated by marching northwards under Sāban-tar, who took Lakhnor, besieged Lakhnautī, but only raised the siege on the arrival of reinforcements from Oudh and the Doāb; and between 1247 and 1258 there were three battles between the Oriyās under the same leader and the Muhammadan forces under Malik Ikhtiyār-ud-dīn Yuzbak-i-Tughril Khān of Bengal. In the last of these battles the latter was defeated, but next year he again led his army to the south and captured and sacked the capital, Umurdan. All this fighting took place during the reign of Nara Sinha Deva, who is, however, better known to posterity as the founder of the beautiful temple of Konarak. The object of most of the raids was to secure the elephants for which Jājnagar, as the Muhammadan chronicles styled Orissa, was famous. The foray of the Bengal Governor, Tughril Khān, in 1279 or 1280 resulted in the capture of a great number of these animals; in 1323 Ulugh Khān, the son of the Delhi Sultān, Ghiās-ud-dīn Tughlik, took away 40 of them; and similar results followed the inroads of the Bāhmani Sultān, Firoz, in 1412, and of Hushan-ud-dīn Hoshang, the King of Mālwā in 1422. The most remarkable of all, however, was the invasion of the Delhi Emperor, Firoz Shāh, in 1360-61. Leaving the baggage behind, the Emperor marched on to Bihār, and then advanced rapidly through the jungles to Orissa. Crossing the Mahānadī, he occupied the royal residence at Banārasi or Cuttack, from which the Rai had fled to an island in the river. Here Firoz Shāh spent several days hunting elephants, and, when the Oriyā king sent envoys to sue for peace, ironically replied that he had only come to hunt elephants and was surprised that, instead of welcoming him, the Rai had taken flight. Finally, the latter sent a present of 20 elephants and agreed to send a certain number annually as tribute, and the Emperor then started on his return journey. It was a disastrous march; the guides lost their way, the army climbed mountain after mountain without finding any road, and it was not till after six months that the exhausted soldiers succeeded in making their way into open country.

The Solar dynasty.

In the meantime, the Vijayanagara kings rose to power, and Orissa was exposed to attack from the south no less than from the north. In 1356-57 Sangama II, the nephew of Bukka I, is said to have defeated the Gajapatis of Orissa; and the Portuguese chronicler, Nuniz, in his account of Vijayanagara (*circa* 1535-37) recorded a tradition that Bukka Rao conquered the kingdom of Orissa. On the death of the last Ganga king, his minister, Kapilendradeva, aided by the nobles and the Bāhmani Emperor, Ahmad Shāh II, seized the throne and founded the Sūryavansa or Solar dynasty in 1435. During his reign of 35 years he was constantly at war. He found the kingdom at a very low ebb, but succeeded by constant wars in extending its limits till it stretched from the Ganges to the Pennar. In Bengal Shams-ud-dīn Ahmad

Shāh was striving to keep up a tottering throne, and here the Oriyās extended their frontier up to the Ganges. In the south the kingdom of Wārāngal had been overthrown, leaving Telingānā divided among a number of petty chiefs, and Kapilendra overran and annexed the country as far as the Krishnā. South of this river, the last two kings of the first Vijayanagara dynasty, harassed by internal revolt and bloody wars with the Bāhmani Sultāns, were struggling to uphold a sinking empire. Taking advantage of their troubles, the Oriyā king annexed the east coast south of the Krishnā as far as Udayagiri near Nellore, and then successfully resisted the attempts of the Bāhmani Sultāns to crush him. In 1457 he forced their army to retire from the siege of Davarakonda, and four years later, on the death of Humāyūn, ravaged their territories up to Bidar. Energetic as was his foreign policy, he showed no less vigour in his internal administration. One of the earliest measures of his reign was to remit the *chaukīdāri* tax paid by Brāhmans and the tax on salt and cowries, to stop the resumption of waste and pasture lands, and to issue orders that all the chiefs in Orissa were to work for the general good on pain of banishment and confiscation of their property. He richly endowed the temple of Jagannāth at Purī, and in this district one of his ministers, Gopīnāth Mahāpātra, built the temple of Jagannāth at Gopīnāthpura about 1465 A.D.

On the death of Kapilendra in 1470, a civil war ensued, each of his sons disputing the throne, but finally Purusottamadeva overcame his rivals with the help of Muhammad Shāh II, to whom he ceded the southern districts of Kondāpalli and Rājāmahendri. His subsequent attempt to recover them led to an invasion by Muhammad Shāh, but the Oriyā king appears ultimately to have regained them, and to have extended his kingdom at least as far as Kondavidu to the south. He also took the opportunity afforded by the confusion which prevailed on the overthrow of the Vijayanagara dynasty to invade that country, and retired with a magnificent booty including the image of Sākshigopāla, which is now at Satyabādī in the Purī district. His son Pratāparudradeva ascended the throne in 1497, and had at once to march to the north to repel an invading army sent by the king of Bengal, Alā-ud-dīn; and 10 years later he had again to drive out another force which advanced under the Bengal general, Ismail Khān. In the south he was engaged in constant wars with Narasa, the founder of the second Vijayanagara dynasty, and with his famous son, Krishnarāya, the struggle ending with the cession of all the territory south of the Krishnā by the Oriyā king. His kingdom was still further reduced by the loss of the tract between the Krishnā and Godāvari in 1522, when Kulī Kutab Shāh, the founder of the Golconda dynasty, invaded Telingānā and drove out the Oriyā army. Although the reign of Pratāparudradeva was one of decline, it witnessed a great religious revival, owing to the spread of the Vishnuite doctrines. In 1510 Chaitanya, the great apostle of Vaishnavism, repaired to Orissa, and there devoted the rest of his days to the propagation of the faith; he is said to have converted the king and several of his officers, but his preaching was not confined to the court, and the purity of his life and doctrines made a lasting impression on the people generally.

The Bhoi dynasty.

The Solar dynasty did not long survive the death of Pratāparudradeva. The powerful minister, Govinda Bidyādhara, killed his two sons one after the other, and in 1541-42 seized the throne. The short-lived Bhoi dynasty which he established only lasted till 1560, and the few years it covered were spent in civil war. First Raghubhanja, the nephew of Govinda, revolted, but was soon defeated and driven out of the country by his uncle. On the death of his son, whose unpopular reign ended about 1557, the minister, Mukunda Deva, rebelled, and after killing the two last Bhoi kings and defeating Raghubhanja, who had returned at the head of a Bengal army, secured the throne in 1560.

Mukunda Deva.

Mukunda Deva, who was a Telugu by birth, was the last independent Hindu king of Orissa, which at this time was in danger from its powerful neighbours both on the north and south. In 1564 Ibrāhim, the Golconda king, was eager for aggrandizement, and in Bengal Sulaimān Karānī was equally anxious to extend his dominions by annexing Orissa. In 1564-65 Mukunda Deva concluded a treaty with the Emperor Akbar, which was intended as a counterpoise to the ambition of the Afghāns in Bengal, but this measure did not long help the Oriyā king. In 1567 Ibrāhim, who had invaded Rājāmahendri unsuccessfully three years previously, conquered the country as far north as Chicācole; and next year Sulaimān Karānī finding Akbar fully occupied by wars in the west, attacked Mukunda Deva when he had marched to the banks of the Ganges, and forced him to take refuge in the frontier fort of Kotsamā. He then detached a part of his force under his Afghān general, Illāhābād Kālā Pāhār, who quickly marched southwards through Mayūrbhanj, defeated the king's deputy, and ravaged Orissa. At this juncture, one of the Oriyā chiefs raised the standard of revolt, and hearing of this, Mukunda Deva hurried south to save his kingdom, but was defeated and slain by the rebel forces, whose leader was in his turn killed by the Muhammadan invaders. Raghubhanja escaped from the prison in which he had been confined by Mukunda Deva, and attempted to secure the empty throne, but after some four months desultory fighting, his death left the Afghāns masters of Orissa (1568 A.D.).

General condition of the country.

Of the internal state of the country during these five centuries of Hindu sovereignty, we have unfortunately very little record. Both Buddhism and Jainism were neglected by the Ganga and the Solar kings, and, if the palm-leaf records can be believed, the followers of those religions were persecuted by the former line. The Gangas did not however neglect the older Saiva worship; the temples of Megheswar at Bhubaneswar and of Chāteswar in Cuttack were built during their rule; and, though they did not build any temples themselves, their rich gifts to the shrines at Bhubaneswar shew that they were the royal patrons of Saivism. At the same time, they seem to have been catholic in their religious tastes, as the great fane of Jagannāth at Purī, the massive sun-temple of Konārak, and probably also the fine temple of Vishnu at Mādhab in this district were built under their orders. The Sūryavansa kings followed in their footsteps, and liberally endowed the Purī temple, and a minister of theirs erected the fine temple of Jagannāth at Gopīnāthpura.

The land was a land of plenty, producing abundance of grain and fruit, and according to Shams-i-Sirāj Afif it was currently reported that 2 *jitals* was the price of a horse, and as for cattle no one would buy them. In spite of this plenty, the people were occasionally exposed to the horrors of famine. The palm-leaf chronicles mention one such famine in the reign of Kapilendradeva when the price of a *bharan* of paddy rose to 105 *kāhāns* of cowries, and another in the same reign when it rose to 105 *kāhāns*, while in the reign of Pratāparudradeva it was once as high as 125 *kāhāns*. Except in times of distress, provisions were exceedingly cheap, cowrie-shells were the only medium of exchange among the people generally, and there was no demand for a gold or silver currency.

Literature and the fine arts were cultivated with some success. Standard rhetorical works were produced at the end of the 13th century; several poems date back to the days of the Solar dynasty; while besides these there were works on law, domestic ritual, astrology and even music. These were written in Sanskrit, but the vernacular was steadily gaining popularity, and the earliest Oriyā works were composed during the latter part of the Sūryavansa rule, such as the Bhāgavata, the great religious work of Jagannātha Dāsa.

The Afghān conqueror was not content, like previous invaders, with levying a ransom from the Province, but marched through it to its southern extremity and besieged and captured Purī. In the year following the conquest, the Afghān king took his departure from Orissa, leaving the government of the country in the hands of a deputy. No sooner was his back turned, however, than the Orissa feudal militia gathered its fragments together for another struggle, and revolted. The Bengal king immediately marched southwards with his Afghān veterans, and succeeded in restoring his supremacy; but he contented himself till the end of his reign in 1572-73 with a distant sway.

THE MUHAMMADAN CONQUEST.

His second son, Dāūd Khān, who succeeded to the governorship of Bengal, threw off all allegiance to the Mughal Emperor at Delhi, and declared himself independent. In the struggle which ensued, the Afghān king was worsted and retired into Orissa. Early in 1574 a great battle took place at Mughalmārī in the Midnapore district, between the Mughals under Munim Khān and Rājā Todar Mal, and the Afghāns under Dāūd Khān, in which the latter were completely defeated. After the battle, Munim advanced upon Cuttack, where a peace was concluded, Dāūd renouncing all claim to Bengal and Bihār, in return for which he received the Province of Orissa as a fief from the Mughal Emperor. Upon the death of Munim Khān, however, in the following year, Dāūd revolted and overran Bengal with his troops. The Afghāns were again defeated in 1576; Dāūd Khān was slain; and two years later, Orissa became a Province of Akbar's empire.

ANNEXATION OF ORISSA BY THE MUGHALS.

The Mughals owed the annexation of Orissa to Akbar's famous Hindu general, Rājā Todar Mal. No sooner had he left Orissa, however, than the Afghān remnant sallied forth from the hill retreats in which they had taken refuge, and in 1580 the Province again revolted against the Empire. Some years of confused fighting followed; and it was not till Akbar sent another Hindu general, Rājā Mān Singh, against Orissa, that any sort of settled government could be restored. Mān Singh completely defeated the rebels in a great battle, captured all forts and strongholds, and finally reannexed it to the rent-roll of the Empire in 1592. From that year the imperial commissions (*sanads*) appointing a Governor of the Lower Provinces regularly include 'Bengal, Bihār, and Orissa.' The Hindu element remained loyal amid the perfidy of the Afghāns; and the representative of the ancient native dynasty, with three of his family, were made grandees of the Delhi Court. Hereafter, the Orissa Afghāns, although they fired up from time to time, found themselves crushed between the Mughal Province of Bengal on the north, and the loyal Hindu dependency of Orissa on the south. In 1598 they took advantage of the Bengal Governor's absence to rebel again, but received so severe a punishment as to prevent any revolt for the next thirteen years. Another rising followed in 1611, which ended in their almost total extermination by the victorious Mughal general. This defeat virtually ended the struggle between the Afghāns and Mughals, and Orissa remained simply a Province of the Mughal Empire until 1751, when the Marāthās obtained it. The remnants of the Afghāns still used it as a basis for marauding expeditions, one of which in 1695-98 attained the dignity of a revolt, and temporarily wrested Western Bengal and Orissa from the Empire. Some time before this the English had appeared on the scene. In 1633 a small expedition headed by Cartwright came from Masulipatam to Orissa and obtained permission to trade in the country. But, so far as this district is concerned, their stay was very short. A factory was founded at Hariharpur, but was abandoned after a few years, and the English settlement was withdrawn.

Orissa, even after the extirpation of the Afghāns, still remained a source of weakness rather than of strength to the Empire. The politic governor who ruled Bengal from 1704 to 1725, Murshid Kulī Khān, in despair of being able to get in its revenues by civil administrators, made it over to soldiers of fortune, who collected the land tax at the spear point, and kept back as much of it as they dared from their distant master. As the latter strengthened his power, however, he sent his son-in-law to govern Orissa in 1706, and annexed the northern part of the Province (now Midnapore district) to Bengal. During the thirty years which succeeded his death, the internal troubles which beset the Mughal Government prevented anything like a settled government in Orissa; the peasantry were left at the mercy of a succession of rude soldiers, who harried the Province and got together as much plunder as their brief tenure of office allowed them.

CESSION OF ORISSA TO THE MARATHAS.

In 1742 the Marāthās came down upon Bengal, and found Orissa an admirable basis for their annual inroads, exactly as the Afghāns had for their revolts. Nine years later, in 1751, the Governor of Bengal, Alī Vardi Khān, bought them off, by practically ceding to them the Province of Orissa, and agreeing to pay twelve lakhs of rupees as *chauth* for Bengal. The treaty of 1751, which severed Orissa from the Mughal Empire, nominally preserved the dignity of the Emperor, and a Musalmān chief was appointed to govern in his name. But although the commissions still bore the Imperial seal, the Emperor's deputy collected the land tax with Marāthā troopers, and made it over to the Marāthā prince. In a very short time this last pageant of dependence upon the Empire disappeared. The Muhammadan deputy of the Emperor was assassinated, and his successor speedily found himself unable to carry on the appearance of a government. The ancient feudal organization among the peasantry and native chiefs, although long since powerless for purposes of defence, still availed for harassing resistance. In 1755-56 the nominal deputy of the Mughal Emperor could not even wring the stipulated Marāthā tribute out of the Province, and begged to be released from his office. A few months later, a Marāthā obtained the undisguised governorship, and from that date till 1803 Orissa remained a Marāthā Province.

THE MARATHA RULE.

Wretched as the state of Orissa had been under the Mughals, a half-century of deeper misery remained for it under the Marāthās. The Marāthā prince had his capital or standing camp at Nāgpur in Central India, and waged incessant war upon his neighbours. His deputies, who were constantly changed, and imprisoned on their recall, struggled to wring out of Orissa—the only peaceful Province of his kingdom—a sufficiency to supply the military necessities of their master. All the offices connected with raising the revenue were sold to the highest bidder at the Marāthā Court at Nāgpur. Every deputy who came to Orissa had ruined himself in order to buy his appointment, and he well knew that the time allowed him for rebuilding his fortunes would be but short. From the hereditary Orissa Prince he managed to wring about £130,000 a year; the smaller proprietors he ousted without mercy from their lands; and he laid heavy burdens upon the pilgrims of Jagannāth. By degrees these atrocities began to work their own cure. The peasant militia of Orissa, strong in the network of rivers, defied the Marāthā troops; and the collection of the revenue in the hilly frontier simply reduced itself to an annual campaign, 'in which,' says Mr. Stirling, 'to say nothing of the expenditure of blood and treasure, the Marāthās were nearly as often worsted as successful.'

There appears to be no trace of anything like a settled administration. The Marāthā cavalry harried the country at stated periods each year, and departed with the spoil; and the internal organization of the village communes formed the only sort of civil government. Each village had its semi-hereditary, semi-elective head, who ruled the hamlet, and represented it to the Marāthā receiver. When the extortions of the latter passed all bounds, the village temporized till it could get its headman out of his clutches, and then the whole community decamped with their cattle into the jungle. But though the swamps and forests yielded an asylum from the Marāthā spearmen, the peasantry could not fly from the consequences of their own flight. The Province lay untilled, and any failure of the rice crops produced a famine. Within seven years two terrible scarcities afflicted Orissa. The famine of 1770, a scarcity of much greater intensity than that of 1866, instead of being mitigated by State importations and relief depôts, was intensified by a mutiny of foreign troops. While the people were dying by hundreds of thousands on every roadside, the Marāthā soldiery threw up the last vestige of control, and for many months ranged like wild beasts across the country. Seven years afterwards, in 1777, another great famine ensued; and as the Marāthā power at Nāgpur decayed, each party into which it split separately harried and plundered the Province.

There were no courts or jails, and the country was infested by thieves and dacoits. If an Oriyā caught a thief in his house at night, he used to brand him by burning, and then let him loose; but sometimes the villagers would rise and kill the thief outright. The Governor's camp-followers lived by plunder, and men struggled to get even this mean post, while to be one of his regular sepoys was to be a king. The Marāthās systematically stripped all rich travellers on the road, while those who escaped the Marāthās were attacked and sometimes killed by the bands of dacoits which lurked in the jungles. Poor people never thought of going to Purī unless they were very pious; those devotees that did always travelled in large bands for mutual protection; and rich men were obliged to retain a strong escort of soldiers armed with swords, spears and matchlocks. The revenue was collected by means of torture and violence. If the people did not pay, they were first beaten with sticks, and then tortured, and in their search for money the Marāthās would dig up the floors, probe the walls, and sometimes pull them down altogether. A favourite mode of torture was to thrust a brass nail between the finger-nails and the flesh, and another was the *chapunī*. This consisted of throwing the man on the ground, placing two crossed bamboos over his chest, and gradually pressing on them till the man consented to pay what was demanded. If he still refused to pay, the operation was repeated on his stomach, back, legs, arms, etc. If the Marāthās saw a man was fat, they said that he had eaten plenty of *ghī*, and must be wealthy—so all people tried to keep lean. If they saw any one wearing clean clothes, they declared he could afford to pay—so all people went about in dirty clothes. If they saw a man with a door to his house, they said it was plain he had money—

so people either did not keep doors, or hid them when the Marāthās underlings were coming. Above all, if a man lived in a masonry house, he was sure to be fleeced, as the Marāthās held that a man who could build such a house could always afford them Rs. 100. They also had another test to find out whether a man had money. They got together the leaves which serve as plates, and on which is served the family repast, and poured water over them; if this did not cover every part of the leaves, they declared that they were greasy, and that the family were all *ghī*-eaters, and must be possessed of money.

THE ENGLISH CONQUEST.

From this terrible oppression the people were delivered by the occupation of the country by the English in 1803. The conquest of Orissa by the English formed a part of the great campaign against the Marāthās in Central India, undertaken by the Marquis of Wellesley. The force destined for the expedition against Cuttack started from Ganjām on the 8th September 1803, under the command of Colonel Harcourt, and marched along the narrow strip of coast between the sea and the Chilka lake. Mānikpatnā was reached on the 15th, having been abandoned by the enemy without resistance. It took two days to cross the dangerous channel through which the Chilka communicates with the sea; and had the enemy made a determined stand there, our position would have been one of considerable danger and difficulty. Leaving Narsinghpatnā on the 18th, our forces entered Purī without opposition. After a halt of two days in the holy city, Colonel Harcourt told off a detachment of Hindu sepoys for the protection of the temple, and resumed his march. The Marāthās, who had gathered in a camp on the other side of the river which flows past the city, at first opened a sharp fire upon our troops, but soon broke and fled. We crossed the river, driving them out of the wood in which they had entrenched themselves. The real difficulties of the expedition now began. There were no roads; the cart tracks, which did duty as roads, were rendered almost impassable by water and mud, and it was with the greatest difficulty that the guns and supplies could be dragged along. The enemy, though not daring to come to close quarters, threw out skirmishers and impeded the progress of our troops by every means which their superior knowledge of the country put in their power. A night attack on the Marāthā camp was made on the 2nd October; the enemy were found leisurely eating their dinner, and were driven out. The Marāthās then took up a position before a town called Mukundpur, near Piplī. On the 4th October they attacked our advanced guard in vastly superior numbers, but were repulsed with considerable loss.

They made good their retreat into the jungles of Khurdā; and no further opposition was offered to the march of our troops, who reached the banks of the Kātjurī a few days after the action at Mukundpur.

The crossing of the river was effected safely; and on the 8th October Colonel Harcourt entered Cuttack city by way of the Lālbāgh, quite unopposed, the gates being open and all the houses empty. The inhabitants had fled in alarm to Tānghī, ten miles north of the Mahānadī, and did not return until the proclamation issued by Colonel Harcourt and Mr. Melvill, C.S., "the Commissioners for settling the affairs of Cuttack," inspired them with confidence in the new rule. Their fears were probably aroused by the restrictions which it was deemed necessary to impose on their personal liberty, and which were not completely removed until November 1805. Had the inhabitants been hostile to our cause and attacked our rear, or fired on our troops from the houses as they marched through the town to storm the fort, the position would have been a critical one. Every precaution having been taken to guard against any such contingency, preparations for the storming of the fort were at once commenced. Six days sufficed to erect the batteries and make the approaches, and the fort was taken by storm on the 14th October.

Equal success attended the expedition against the town of Balasore, which had been despatched from Bengal; and the three principal towns of the Province having fallen into our hands, a part of the force was, in pursuance of the original plan of the campaign, despatched under Major Forbes to force the Barmūl Pass. Colonel Harcourt with another detachment marched against Kujang, by way of Patāmundai. The Rājā of Kujang had been detected carrying on a correspondence with the Rājās of Kanikā and Harishpur, with a view of entering into a triple alliance, offensive and defensive, against the British authority. The Rājā fled as soon as he received tidings of the near approach of the troops. His elder brother, whom he had kept a close prisoner, was released and placed on the *gadi*, and a large reward was offered for the apprehension of the fugitive, who was captured shortly afterwards and confined in the fort at Cuttack. His fortifications were all dismantled, and the cannon found in them carried away to Cuttack. Before returning, Colonel Harcourt completed the success of his expedition by reducing to submission the turbulent Rājās of Kanikā and Harispur. Their forts were also demolished, and the guns found in them taken away. In carrying out these measures no resistance was met with; and they were undertaken more with a view of impressing the people with a sense of the strength of the British arms than from the necessity of putting down any serious armed opposition.

THE KHURDA REBELLION.

The conquest was effectual and complete, and the district has since enjoyed a tranquillity broken only by the Khurdā rebellion of 1817. The *paiks*, or old landed militia, broke out into revolt, in consequence of the ruin and oppression which the early system of English government brought to them. Owing to the resumption of their service tenures, they had been deprived of the lands which they had enjoyed from time immemorial, and were subjected to the grossest extortion and oppression at the hands of the farmers, *sarbarāhkārs* and other underlings, to whom our Government entrusted the collection of the revenue, as well as to the tyrannies of a corrupt and venal police. They formed a wild and motley crowd with their war dress of a cap and vest made of a tiger or leopard skin, a sort of chain armour for the body and thigh, and a girdle formed of the tail of some wild animal; and they further heightened the ferocity of their appearance by smearing their limbs with yellow clay and their faces with vermilion. But savage though their equipment was, they fought well, defeating the troops first sent against them, and for some time held undisputed possession of the country. The insurrection was most formidable in the sub-division of Khurdā and that part of the Purī district immediately adjoining it, but the rising was pretty general all over the southern and eastern parts of Orissa. In Cuttack the *paiks* burnt the thānas of Asureswar, Tiran and Hariharpur, and committed various ravages in the country surrounding them. They had however no recognized leader of ability in these parts, though they were secretly encouraged by the Rājās of Kujang and Kanikā; and their actions were more those of bands of dacoits than of a people striving to rid themselves of a foreign yoke. An expedition was despatched against them from Cuttack, a force of about 2,000 *paiks* was utterly routed near Kujang, and the Rājā, seeing that further resistance was hopeless, surrendered. By the end of October 1817 British authority was completely restored in Kujang, the detachment stationed there was withdrawn, and military law ceased to be in force. Though bands of *paiks* continued for some time to infest the jungles of Khurdā, the rebellion in this district was completely stamped out, and the country gradually became pacified and soon recovered its accustomed tranquillity and security.

CHAPTER III.

THE PEOPLE.

GROWTH OF POPULATION.

Early estimates.

PRIOR to 1872 no regular census of the district by the simultaneous enumeration of the people had ever been taken; but several rough attempts were made from time to time to estimate the number of inhabitants. Stirling in his Account of Orissa, written in 1822, estimated the population of the district as 1,296,365 persons, his calculations being based upon an enumeration of the dwellings, allowing 5 persons to each house. Twenty years later the Revenue Survey of 1842 returned the population of Cuttack at 553,073 ; a subsequent attempt at a census shewed it as 800,000; and yet another estimate was made in 1847, according to which the district contained 1,018,979 persons, giving an average pressure of 320 persons per square mile. All these estimates were very rough, as they were made simply by counting the houses through the agency of the police and assigning an average number of inhabitants to each dwelling. In 1855, however, an attempt was made to obtain more accurate figures, special officers being appointed to test the returns by counting the houses and their actual inhabitants in different parts of the district, and thus to ascertain the correct average for each house. The result disclosed an average of slightly over 5½ inhabitants to each dwelling, and the population was returned at 1,293,084. About ten years afterwards another rough census was taken at the close of the famine of 1866, the landholders being called upon to submit returns of the surviving inhabitants; and it was estimated that the total population amounted to 1,072,463 persons.

Census of 1782, 1881 and 1891.

The first census which can be regarded as in any way approximating to the truth was taken in 1872, by which time the population had probably increased materially owing to the return of the people who had fled from their homes during this last great calamity. That census disclosed a total population of 1,494,784 souls with an average density of 470 to the square mile; and in 1881 it was found that the population had risen to 1,738,165 and that the pressure of the population amounted to 494 persons per square mile. The census of 1891 showed a large increase, the total number of persons recorded being 1,937,671 and the density 533 per square mile. A portion of the increase was due to the annexation of Bānki with a population of 57,368 and an area of 116 square miles; but even if the figures for that tract are included in the previous returns, the growth of population was 16·24 per cent. between 1872 and 1881 and 7·9 per cent. in the decade ending in 1891. The census of 1872 however was probably incorrect, and it has been suggested that the actual population exceeded the census figures by at least 100,000. This would reduce the rate of growth in the succeeding nine years to 13 per cent., which is about what might be expected during the period when the district was recovering from the terrible famine of 1866. The progress during the next decade would probably have been greater, were it not that the district suffered generally from repeated outbreaks of cholera, and that in certain localities scarcity and the great cyclone of 1885 seriously affected the growth of the population. In the head-quarters subdivision, Bānki remained almost stationary owing to a scarcity almost amounting to famine, which, although it caused no deaths, drove a large number to emigrate. In the Kendrāpāra sub-division the great cyclone of September 1885 destroyed 45 villages in thāna Patāmundai, most of the inhabitants of which were either drowned or succumbed to the fever and cholera which usually form the sequel of such calamities, while those that survived emigrated to tracts less exposed to the destructive action of storm-waves.

Census of 1901.

The result of the census of 1901 was a further increase of 125,087, or 6·5 per cent., the diminution of the rate of growth as compared with that in the previous decade being probably due to the loss suffered by the movements of the people. The general increment was shared by all parts of the district, and the rate of development was remarkably uniform throughout. The growth of population was least in the already densely inhabited thānas, Cuttack, Sālīpur, Jājpur and Jagatsinghpur, and greatest in the sparsely inhabited thānas, Patāmundai and Aul on the sea-coast, where the construction of protective embankments and the offer of easy terms of settlement led to considerable reclamations of land which had been thrown out of cultivation by the salt-water floods of 1885. The inland thānas adjoining the Garjāt States, Dharmshāla and Bānki, which have a very sparse population, came next; and then Kendrāpāra, where the population, though more dense than in the western part of the district, is less so than in the central thānas.

Density. Density of population is very largely determined by the physical aspects of the three distinct areas into which the district is divided. In the maritime police circles the pressure of the population, which falls in the Aul thāna to 298 persons to the square mile, is greatly reduced by a belt of saline soil, in places as much as 30 miles wide, running along the sea and covered by sand, coarse grass or shrub, in which agriculture is almost unknown. In the submontane strip, a region of rocky hill and barren soil, which supports a scanty and semi-Hinduized population, the density is somewhat higher than in the salt tract, but is still comparatively low, being 366 persons to the square mile in the Dharmshāla police circle and 377 in Bānki. The alluvial plain lying between these two estimates is highly cultivated, and has in parts a density of population very little less than that of the most thickly inhabited parts of Eastern Bengal and Tirhut. The pressure is greatest (933 persons to the square mile) in the Sālīpur thāna, lying in the heart of the district between the two principal branches of the Mahānadī river, where nearly every field is reached by the canals and distributaries of the great Orissa irrigation system. In the neighbouring thānas of Jājpur and Kendrāpāra the land bears 842 and 765 persons to the square mile; and considering that the population is almost entirely agricultural, the density may reasonably be considered to be very great. Taking the district as a whole, the density per cultivated square mile is 1,084, the cultivated area, according to the returns of 1903-04, being 1,903 square miles. On the other hand, it has been calculated that each square mile of unirrigated land would support 1,167 persons and each square mile of irrigated land 1,515 persons; and there is, therefore, still room for the expansion of the people on the soil which has been brought under the plough, though the density is already very great.

Migration. From the fact that in the census of 1901 the number of persons born in Cuttack who were enumerated elsewhere in India was 116,759, and that only 32,944 persons born elsewhere were enumerated in this district, it will be apparent that the number of emigrants is far in excess of the number of immigrants. Large numbers are attracted to the sparsely inhabited Native States forming the western boundary of the Division, where much arable land is still unoccupied; and to judge from the equality of the sexes, the emigration into the contiguous districts is also probably permanent. The net excess of emigrants to the adjoining districts of Balasore and Purī is about 14,000—a result which is only to be expected, as Cuttack is the most densely populated district in the Division. It loses considerably more than this number, however, by migration to distant places, chiefly to the metropolitan area, Assam and the Central Provinces. Large numbers go to Calcutta and its neighbourhood to serve as *pālki*-bearers, *darwāns* and labourers; natives of the district are found working as cooks and domestic servants throughout Bengal; and numerous emigrants go to the Sundarbans as cultivators and field labourers. This overflow is, however, mostly temporary or periodic, and its most noticeable feature is the very small proportion of women who accompany the men, only 4,179 of the 54,197 emigrants enumerated in distant parts of the Provinces being women. The advent of the railway has naturally afforded far greater facilities for communication with the outside world than previously existed, and has greatly stimulated migration. Natives of the district employed in Bengal return home at much more frequent intervals than formerly, and, on the other hand, the number seeking employment elsewhere has greatly increased. Of the immigrants, the greater portion (22,664) come from the adjoining districts, and the remainder is almost entirely made up of immigrants from other Provinces. The number of the latter (8,161) was however swelled by the sepoys belonging to a wing of a Madrāsi regiment which was stationed at Cuttack at the time of the census.

Towns and villages. There are only three towns, Cuttack, Jājpur and Kendrāpāra, with a population of over 5,000, and the total number of their inhabitants is only 78,720, or 3·8 per cent. of the population. The remainder of the people are clustered together in 5,517 villages. The people have hitherto developed no tendency towards city life; and while the rural population has grown very largely during the last 30 years, the towns have done little more than hold their own. No new centres of industry have sprung up, and the rapid development of commerce and manufactures which is so powerful a factor in the increase of urban population is as yet unknown. The total number of towns-folk has increased only by 12,000 in the last 30 years, and the two towns of Jājpur and Kendrāpāra contain only 3,330 more inhabitants than they did in 1872. Even Cuttack, the capital of Orissa, has shown little progress as a city, though it focuses the trade of that Province. Situated at the first bifurcation of the river Mahānadī, protected by massive embankments from its floods, and forming the nucleus of a widely ramified system of canals, it nevertheless has failed as yet to attract the homestead-loving people of Orissa in any large numbers; and though it is the largest town in Orissa, its inhabitants still number only 51,364. Of late, however, there appears to have been a greater influx of permanent settlers, owing to the advent of the railway, and its population has grown by 9 per cent. since 1891, while the district, taken as a whole, has added only 6·5 per cent. to its numbers. The Oriyā appears to have an inherent aversion to town life; he will not voluntarily leave his hereditary fields, and even when forced to betake himself to a town, he strives to reproduce his village life in his new surroundings. The distinction generally between an urban and rural population is primarily in respect of occupation, the agricultural class naturally predominating in the villages, while in the towns, where the trading and professional classes form the majority of the population, it is an unimportant section. Except in Cuttack, however, the distinction is not by any means well marked, as Jājpur and Kendrāpāra comprise a number of more or less scattered hamlets, the inhabitants of which are to a greater or less extent employed in agriculture; and even in Cuttack with its crowded streets and bazars, many parts are distinctly rural in character with trim homesteads nestling in small orchards.

Sex and age. In common with the other districts of Orissa, Cuttack has a marked excess of females over males, there being 1,072 females to every thousand males. The two local castes of fairly high status (Karan and Khandait) have a far larger proportion of women than those of equal rank elsewhere, and among the functional groups the excess of females is greater than anywhere else in Bengal. The proportion of unmarried persons is also higher than in other parts of Orissa, viz., 525 out of 1,000 males and 348 out of every thousand females.

The number of children under 10 per 1,000 of the population has fallen considerably since 1881, and the proportion they bear to the number of married women aged 15 to 40 is now lower than in any other part of Bengal, except South Bihār. This decline may be attributed to the fact that in 1881 the district was recovering from the great famine of 1866, *i.e.*, its population was growing at a specially rapid rate, and the proportion of young people was therefore exceptionally high. The population has now regained its normal condition, and the proportion of children has fallen accordingly. The average age of the people has been steadily increasing during the last 20 years; the population has grown by more than the Provincial average in spite of the low proportion of children; and it may be concluded that the people enjoy a longer span of life than those in many other parts of Bengal.

LANGUAGE. Oriyā is the mother-tongue of the large majority of the people, but English, Hindī, Bengali and Telugu are also spoken. English is the language of the small English settlement, of the larger Eurasian element and of the better educated natives. Hindī is used by a large number of the Muhammadan residents of the district, by members of the police force who have been recruited from up-country, and by the pensioned sepoys of the various Madras regiments which have garrisoned Cuttack. There are always a certain number of Bengalis among the professional classes in the district; and Telugu is spoken by some weavers, sweepers and others who came and settled here during the last Madras famine, as well as by other immigrants from that Presidency: it is sufficiently common to make it necessary to employ an interpreter in the Criminal Courts at Cuttack.

Oriyā language.

More than 96 per cent. of the people speak Oriyā,* or as it is sometimes called Odrī or Utkalī, *i.e.*, the language of Odra or Utkal, both of which are ancient names for the country now called Orissa. Oriyā, with Bengali, Bihāri and Assamese, forms one of the four speeches which together make up the eastern group of the Indo-Aryan languages. Its grammatical construction closely resembles that of Bengali, but it has one great advantage over Bengali in the fact that, as a rule, it is pronounced as it is spelt. There are few of those slurred consonants and broken vowels which make Bengali so difficult to the foreigner. Each letter in each word is clearly sounded, and it has been well described as "comprehensive and poetical, with a pleasant sounding and musical intonation, and by no means difficult to acquire and master." The Oriyā verbal system is at once simple and complete. It has a long array of tenses, but the whole is so logically arranged, and built on so regular a model, that its principles are easily impressed upon the memory. It is particularly noticeable for the very complete set of verbal nouns, present, past, and future. When an Oriyā wishes to express the idea embodied in what in Latin would be called the infinitive, he simply takes the appropriate verbal noun, and declines it in the case which the meaning necessarily requires. As every infinitive must be some oblique case of a verbal noun, it follows that Oriyā grammar does not know the so-called infinitive mood at all. In this respect Oriyā is in an older stage of grammatical development than even classical Sanskrit, and, among Indo-Aryan languages, can only be compared with the ancient Sanskrit spoken in the Vedic times.

The archaic character, both of form and vocabulary, runs through the whole language, and is no doubt accounted for by geographical position. Orissa has ever been an isolated country bounded on the east by the ocean, and on the west by the hilly tracts inhabited by

* This account of the Oriyā language has been condensed from Dr. Grierson's Linguistic Survey of India, vol. V.

wild aboriginal tribes. On the south the language is Dravidian and belongs to an altogether different family, while, on the north, it has seldom had political ties with Bengal. On the other hand, the Oriyās have been a conquered nation. For eight centuries Orissa was subject to the kings of Telinga, and, in modern times, it was for fifty years under the sway of the Bhonslas of Nāgpur, both of whom left deep impressions of their rule upon the country. On the language they imposed a number of Telugu and of Marāthī words and idioms, which still survive. These are, so far as we know, the only foreign elements which have introduced themselves into Oriyā, except the small vocabulary of English court terms, and a few other English expressions, which English domination and education have brought into vogue. Cuttack, especially the town, is however to a certain extent affected by Bengalisms, owing to the residence there of a number of Bengalis who have been settled in the district for some generations. In former times sales of Orissa estates for arrears of land revenue were held in Calcutta, and the purchasers were frequently Calcutta Bengalis who settled in Cuttack. These Bengalis and their descendants have developed a curious jargon of their own, their ancestral language being interlarded with Oriyā and Hindī expressions. Owing to their frequent use of the word *kare*, a corruption of the Oriyā *kari*, their speech is vulgarly known as *kerā* Bengali; and this mongrel language has in its turn reacted on the local Oriyā.

Written character. Oriyā is encumbered with the drawback of an excessively awkward and cumbrous written character. This character is, in its basis, the same as Devanāgari, but is written by the local scribes with a stylus on a talipot palm-leaf. These scratches are, in themselves, legible, but in order to make them more plain, ink is rubbed over the surface of leaf and fills up the furrows which form the letters. The palm-leaf is excessively fragile, and any scratch in the direction of the grain tends to make it split. As a line of writing on the long, narrow leaf is necessarily in the direction of the grain, this peculiarity prohibits the use of the straight top line, or *mātrā*, which is a distinguishing characteristic of the Devanāgari character. For this the Orissa scribe is compelled to substitute a series of curves, which almost surround each letter. It requires remarkably good eyes to read an Oriyā printed book, for the exigencies of the printing press compel the type to be small, and the greater part of each letter is this curve, which is the same in nearly all, while the real soul of the character, by which one is distinguished from another, is hidden in the centre, and is so minute that it is often difficult to see. At first glance, an Oriyā book seems to be all curves, and it takes a second look to notice that there is something inside each.

Literature.

The earliest example of the language which is at present known consists of some Oriyā words in an inscription of king Nara Singha Deva II, dated 1296 A.D. An inscription of Nara Singha Deva IV, dated 1395 A.D., contains several Oriyā sentences, which show that the language was then fully developed, and was little different from the modern form of speech either in spelling or in grammar.

Oriyā literature* is however of comparatively recent growth, none of the existing works, so far as can be ascertained, going back beyond the 16th century A.D. It consists exclusively of verse, and, as is natural with a conservative people like the Oriyās, the earliest works extant are religious, viz., a few songs and certain paraphrases of the Sanskrit Purānas and epics. No work is so much venerated as the *Bhāgavata* of Jagannātha Dāsa; and next in estimation come the *Rāvāna* of Balarāma Dāsa, the *Bhārata* of Sārolā Dāsa and the *Harivansa* of Achyutānanda Dāsa. All these were composed in the first half of the 16th century A.D., to which period may probably be referred popular songs like the *Kesabakoili* or cuckoo-song about Krishna. Profane literature appeared later, and at first dealt only with mythological stories. Among the oldest of these is the poem *Rasa-kallola* by Dīnakrishna Dāsa. This poem describes the early career of Krishna, and is a favourite with the Oriyās; its versification is peculiar in making every line begin with the same letter *ka*. The most famous of the Oriyā poets is Upendra Bhanja, who, following Dīnakrishna Dāsa, flourished in the early beginning of the 18th century. One of the royal family of Gumsur, a petty hill State in the north-west of Ganjām district, he was driven to take refuge in Orissa in the course of a civil war, and there devoted his life to Oriyā literature. Of his voluminous compositions, forty-two are at present known, the bulk of them consisting of poems with love-stories as their theme. He was apparently the first Oriyā poet to free himself from the trammels of exclusively religious and mythological influences. His poems labour under the defects of obscenity and unintelligibility; but they show at the same time a master's hand in letter-selection or rhetorical excellence. In these qualities his only rival is Abhimanyu Sāmanta Singhār, a zamīndār of Golakunda in the Jājpur sub-division of this

* I am indebted to Bābu Monmohan Chakravarti, M.A., B.L., M.R.A.S., for this account of Oriyā literature.

district, who died in 1806 when only 49 years old. His poem *Bidagdha-Chintāmani* is a veritable store-house of rhetorical excellence, while its latter cantos explain in lucid lines the abstruse doctrines of Vaishnavite *Bhakti* and *Prema*.

During the British period Oriyā poetry has shown no progress. It is represented by a few doggerel compositions and some small pieces of verse, among which a collection of short poems by Rai Rādhā Nāth Rai Bahādur, late Inspector of Schools, Orissa, deserves notice. Prose is, however, being carefully studied and has a promising future.

RELIGIONS.

Hindus.

Orissa is the Holy Land of the Hindus, which through all its vicissitudes has held its high place in the religious esteem of the people. The Purānas are full of descriptions of its sanctity, and it is declared to be the favourite abode of the Devatās and to boast a population composed, more than half, of Brāhmans. From end to end, we are told, it is one vast region of pilgrimage (*tīrtha*); its happy inhabitants live secure of a reception into the world of spirits; and those who visit it and bathe in its sacred rivers obtain remission of their sins. According to popular belief, even the victorious Musalmān who led Akbar's invading host into this land of sanctity, was struck with amazement at its sacred river, the Mahānadī, its vast crowds of Brāhmans and its lofty temples of stone, and exclaimed "This country is no fit subject for conquest or for schemes of human ambition. It belongs entirely to the gods and is one great region of pilgrimage throughout." From the moment the pilgrim crosses the Baitaranī river he treads on holy ground; and in these circumstances it is not surprising that the great bulk of the population of the district is composed of Hindus, who with 2,002,573 souls account for 97 per cent. of the people.

Muhammadans.

Practically all the remainder are Muhammadans, who number 57,356 persons, or 2·78 per cent. of the people. It is somewhat strange at first sight that they are not more numerous, considering the strong footing they once had in the district. Badāoni, who was a zealous Moslem, describes Cuttack as being a mine of unbelief when it was subdued by Sulaimān, but the Muhammadans effectually conquered the Province and took complete possession of it in 1568 A.D.; and subsequently in Akbar's reign, when the Afghān kingdom of Bengal was overthrown by the Mughals, the Afghāns migrated in large numbers into Orissa and there held large fiefs and independent power. When they again rose in revolt, they were signally defeated, and in order to deprive them of the means of political combination, Shujait Khān distributed them among the villages in the interior, but allowed them grants of lands sufficient to maintain their dignity. Orissa long remained a dependency of the Mughal Empire, and the Afghāns continued in possession of their *jāgīrs*; but with the lapse of centuries they dwindled in numbers and in influence. The Muhammadan conquest was not only late chronologically, but it failed to attain that permanence and completeness which it obtained in Bengal. It was a conquest rather than a colonization, the Mughals and Afghāns made few converts to Islām, and the present Muhammadan residents of the district are nearly all descendants of the invaders. Their number has increased but little since 1872, when they amounted to 40,262 persons, but on the other hand their growth has been relatively greater than that of the followers of other religions. They form a small community, mostly in easy circumstances—a fact which favours their multiplication at a rate hardly attainable by the general mass of the population.

Christianity was practically the only other religion represented at the census of 1901. The number of Christians (2,652) is still comparatively small; and they are almost entirely confined to Cuttack town, which is the head-quarters of the Baptist and Roman Catholic Missions. The labours of the Baptist missionaries date from 1822, but in spite of their earnest efforts they have made but little progress in actually converting the people. The first native convert was baptized six years after the establishment of the Mission; and by 1872 the total number of native Christians was only 1,911. This growth was to a great extent due to the fact that the famine of 1865-66 contributed largely to its ranks, as over 650 famished children, whose parents had died of starvation or had deserted their offspring in the last extremity of famine, were rescued and brought up in the Christian faith. Since 1872 the Christian community in Cuttack has remained practically stationary; among those who lost caste from eating in the mission kitchens were many old men and women who left no children to take their place; and the total number of native Christians attached to the Mission is even now only about 2,000. On the other hand, it has done an immense amount of indirect good; the Baptist missionaries were the first to start properly conducted schools; and the Cuttack Mission Press, which has the distinction of being the oldest press in Orissa, has sent forth a stream of civilizing literature. In addition to this press, the Mission maintains an English High school affiliated to the Calcutta University, an European High school, and an orphanage for boys and girls. There are in all 8 missionaries and 5 evangelists attached to the Mission in this district. Christians.

The Roman Catholic Mission has been established in the town of Cuttack since 1845, and has a congregation of about 400 Roman Catholics, nearly one half of whom are Europeans or Eurasians, the remainder being natives, chiefly immigrants from the Madras Presidency. The Mission has a convent, a chapel and a church; the latter, which has accommodation for 200 worshippers, was built in 1858-50. It also maintains an orphanage for children of all classes, which was founded in 1880 and contains 60 inmates. The latter are under the direction of 5 sisters sent out by the Congregation of the Sisters of St. Joseph; the priests of the Mission are sent to India by the Congregation of the Missionaries of Saint Francis of Sales.

Hindu sects.

Vaishnavism is predominant among the common people of this district, and the causes of this predominance are not far to seek. The existence of the temple of Jagannāth, who is regarded as the incarnation of Vishnu, has exerted a powerful influence on the popular faith; and besides this the famous reformer Chaitanya passed an important part of his life in these parts, and made a lasting impression upon the popular mind by the purity of his life and teachings. Vaishnavism is still struggling to divert the popular mind from the number of gross animistic accretions by which the religion of the mass of the population is encumbered; and it is Vaishnavism which mainly distinguishes the semi-Hinduized aborigines in the plains of Orissa from their Animistic brethren in the hills, though its adoption is merely nominal and its high ethical principles do not shape the moral conduct of the people. Genuine Oriyās belonging to sects other than that of the Vaishnavas are very few in number. Sāktas, the followers of Sakti or the Goddess Durgā or Kālī, are to be found among the Brāhmans and Kshattriyas of the district and among the Bengali immigrants. The only other two sects represented, the Saivas and Gānapatyas or followers of Ganapati, bear a very small ratio to the total Hindu population. The Parwārs of Rājputāna, who are Jainas, have a temple consecrated to the worship of Parasnāth at Chaudhri-bazar in the heart of Cuttack town.

Religious life.

The religion of the people exhibits very clearly the blending of Hinduism with Animism, and the process of assimilation appears to be illustrated by the common legend of Jagannāth.* Here we find the aboriginal people worshipping a blue stone in the depths of the jungle, until the deity grows tired of the jungle offerings of the primitive people and longs for the cooked food of the more civilized Aryan race. When the Aryan element at

* See Sir W. W. Hunter's Orissa, Vol. I, pp. 88—95.

length comes on the scene, the rude blue stone disappears and gives place to a carved image. At the present time this twofold worship co-exists throughout Orissa. The common people have their shapeless stone or block which they adore with simple rites in the open air; while side by side with it is a temple to one of the Aryan gods with its carved image and elaborate rites. Every village pays homage to the Grām Devatī* or Thākurānī, as these stones and stocks are called, and reverence her as the tutelary goddess of their small community.

The goddess is commonly represented by a piece of shapeless stone, smeared with vermilion and surrounded by several smaller pieces of stone, also vermilion-daubed and shapeless, which represent her children. Carved images are sometimes, though rarely, met with, and occasionally the trunk of some tree, supposed to possess supernatural properties, is smeared with vermilion and worshipped as the village goddess. Besides the generic name Grām Devatī, each goddess has a separate specific name, which is commonly one of the thousand names of the goddess Kālī. The general idea seems to be that she is like a mischievous old witch; and earthen figures of horses, elephants and other animals are placed before her by the superstitious rustics, as it is believed that she wanders about at night.

The most noticeable feature of the Grām Devatī worship is the non-priestly caste of the men who conduct it, the Bhandāri, Māli, Rāul or Bhopā being usually the priest. They hold small rent-free grants called "*māfi Grām Devatī*," *i.e.*, lands which were left unassessed for her worship at the time of the first regular settlement; and they also receive daily doles from the rich men of the village and weekly doles from the poorer people: the latter are given on Thursday, commonly regarded as Lakshmī day, or the day of the goddess of fortune, which is considered a specially auspicious day for the regular *pūjā* of the Grām Devatī. The first essential in this worship is a bath which keeps the Thākurānī cool and well disposed towards the village. The bath includes smearing with *ghī* and turmeric; when it is completed a paint of vermilion is put on, and after the toilet is over a light oblation (*bhog*) of fruits and other sweetmeats is offered. The daily *pūjā*, including both bath and *bhog*, costs about an anna, and if this small daily expenditure cannot be met, the priest contents himself by pouring a little water over the goddess, though sometimes even this inexpensive offering is dispensed with. The

* For a fuller account *see* Note on the Grām Devatī or Tutelary Village Deity of Orissa by Bābu Jāminī Mohan Dās, J. A. S. B., Vol. LXXII, Part iii, No. 2, 1903.

worship of the Grām Devatī is conducted with great pomp and ceremony on the Mahāstamī or second day of the Durgā *Pūjā*, and special offerings of sweetmeats and fruits are made on all festive occasions. The Thākurānī, who is supposed to possess more powers for doing or averting mischief than for doing positive good, receives special attention on the outbreak of any epidemic disease. Within her own village she is believed not to commit any mischief, and epidemics are supposed to be the work of neighbouring goddesses, whom the tutelary village goddess expels by persuasion or superior force, if she is duly propitiated. The occurrence of a single case of cholera in the village is the signal for "Thākurānī Mārjanā" or washing of the Thākurānī. The villagers immediately raise the necessary funds by subscription, and propitiate the goddess by a cooling bath and refreshing offerings, the ceremony being repeated, if the epidemic does not cease.

The people have a peculiar means of knowing the wishes and decrees of the goddess. In almost every village there is a male or female medium, called Kālasī, through whom the goddess communicates with the people. The presentation of a betel-nut is the token of engaging the Kālasī, whose services are specially in demand on the occasion of an outbreak of cholera. Before the time appointed for the Mārjanā, he takes a purifying bath, puts on a new cloth, and paints his forehead with vermilion. Then holding two canes in his hands, he appears before the Grām Devatī, and with dishevelled hair swings his body to and fro. After a time he begins to tremble, and in the course of his confused mutterings gives out some secrets of the village to win the confidence of the people. He then predicts evil to some and good to others, prescribing at the same time the remedies required, which take the shape of offerings to the goddess and special favours to himself. While going through these antics, the Kālasī is sometimes offered a fowl, the blood of which he drinks after pulling off the head.

Certain village goddesses are regarded as "Parama Vaishnavis" or devoted followers of Vishnu, and animal sacrifices are not allowed before them. Probably owing to the spread of Vaishnavism, such sacrifices are only made sparingly before the other goddesses; but in the Mahastamī *pūjā* and other special *pūjās* offered in fulfilment of vows, animals are generally sacrificed. Fowls are also let loose before some of the goddesses by the upper classes of Hindus, and are killed and eaten by the lower classes.

It seems hardly open to question that this worship of the malevolent spirit, through the medium of shapeless stones, is an offshoot of the fetishism of the aborigines. The fact that all Hindus from the highest to the lowest make the Grām Devatī the object of their adoration shows how the beliefs of the whole Hindu community have been permeated by this fetishism. It still includes, though to a restricted extent, the sacrifice of animals, which is one of the most characteristic features of aboriginal worship; and the offering of fowls, which are so rigorously excluded from the houses of the upper classes of Hindus, can hardly be said to be anything else than an aboriginal practice. The restriction of the priestly function to the Sūdra castes is another link in the chain of circumstances which indicate the aboriginal origin of this form of worship. While the Brāhman stood aloof, the mass of the people, leavened in their lower strata by the aborigines, adopted the faith which, by its easy explanation of the origin of evil, appealed most strongly to their simple minds. The Brāhman could not, however, long stand against the popular current which thus set in, and he eventually invented more refined forms of worshipping the same malevolent spirit. The aboriginal mode of village worship seems thus to have preceded the Paurānik rites of Sakti worship, although the present names of the goddesses are apparently of later date.

SOCIAL LIFE. Character of the people.

The Oriyā has long had an unenviable reputation as a weak, effeminate and stupid creature, and early writers almost all condemn them. It is said that they "prefer dirt and scarcity of food with idleness to cleanliness and plenty with hard labour," and that "industry and enterprise are as foreign to them as opium-eating and noon-day sleep to the English husbandmen;" and this unfavourable estimation of their character was forcibly expressed by Stirling, who, writing in 1822, says:—"The Oorias as a nation are justly described by Abul Fazl to be very effeminate, that is, they are extremely deficient in manly spirit, their figures are slight and delicate, and the costume of the males has little to distinguish it from that of the females, except the different manner of wearing the cloth fastened about the loins. They are moreover equally ignorant and stupid. Orissa might be termed the Bœotia of India, with reference to the intellectual dullness of its inhabitants as compared with the people of any other province. A striking proof of the estimation in which their capacity has been ever held is the fact that in all ages and under all Governments since the downfall of the Orissan monarchy, the principal official employments throughout the Province have been engrossed by foreigners—by Bengalis north and Telingas south of the Chilka Lake—owing, I really believe, in a great measure to the difficulty of selecting from its indigenous population persons properly qualified for trusts of difficulty and importance. The mass of the people are little prone to the commission of crimes of a daring and heinous character, as might be inferred from the feminine spirit above ascribed to them; but they are well versed in all the arts of low cunning, dissimulation and subterfuge, and the love of intrigue forms a prominent feature in their character, however clumsy many of their attempts to mislead or circumvent. In justice however to the bulk of the agricultural population, it must be said that the ryots of Cuttack are extremely industrious though they work with little spirit or intelligence, and altogether the Oorias of the plains, whatever their faults, are certainly the most mild, quiet, inoffensive, and easily managed people in the Company's provinces. They furnish too a valuable class of servants in whom the virtues of fidelity and honesty (according to their own conception of those qualities) are conspicuous."

Experience has proved that the character of the people given by Stirling was written in ignorance of the Oriyā character; he was too honest to misrepresent, too just to be prejudiced against them, but coming in contact only with those who under excited passions were engaged in law-suits, he did not know the people. It is true that they are ignorant, superstitious and priest-ridden, and have less natural acuteness than the Bengalis. But they are less prone to litigiousness and deceit, they evince more gratitude for kindness, and they are more impatient under ill-usage. They have proved themselves to possess great aptitude for public business, and their uprightness in offices of much responsibility and beset with very great temptation has placed them in the first rank of our native subjects for trustworthiness and honesty. To this it should be added that the Oriyā bearers are industrious, sober, faithful and trustworthy, and that they will follow a kind master to all parts of India. On the other hand, it is true that the labouring classes and the lower castes of cultivators are somewhat dull-witted, but it is doubtful if the Bihāri peasant or the Bengali ryot is mentally the superior of the Oriyā. Among the more educated classes the Brāhmans are obstinate and bigoted, but they are of a refined and intellectual type; and the writer caste of Karans, or as they are called locally the Mahāntis, have as high a reputation for acuteness as the Kayasths of Bihār. The old reproach of unfitness for Government employment can no longer be levelled against them. The pure foreign element has almost disappeared from among the ministerial establishment; and even the domiciled Bengali who has adopted the country as his own is losing ground before the advancing native of Orissa. Young Oriyā graduates passing out of the Ravenshaw

College at Cuttack are also entering the Subordinate Executive Service; and at the present rate of recruitment, there will in a short time be a staff of Oriyā officers sufficient for the requirements of the whole Division. Generally speaking, the people are kindly and good humoured, easily controlled and remarkably law-abiding; they are given only to committing the pettiest of crimes, and in many ways recall the old idea of the mild Hindu. Outside Orissa they have a reputation for good work as domestic servants, chaprāsis and coolies; and their readiness to migrate and find employment as carpenters, punkah-pullers, *pālki*-bearers, and in other kinds of labour requiring physical rather than mental capacity seems to show that they are not altogether sunk in the stagnation which earlier writers have attributed to them. Even, however, among the literate classes there is a want of enterprise, slowness, and hopeless conservatism which are in marked contrast to the versatility of their Bengali cousins; and an inborn love of ease and dislike of hard work appears to permeate all classes.

In justice, however, to the Oriyās it should be remembered that for ages they have been a conquered nation, and that within the last few centuries they suffered at the hands first of the Mughals and then of Marāthā conquerors. From the end of the 17th century they were continually harried and oppressed; under the Mughal Emperors a greedy and generally disloyal Deputy wrung from the Province an uncertain revenue; the wretched peasantry were ground down beneath a military occupation; and a rapid succession of rude soldiers harried the country and got together as much plunder as they could wring from the people. But miserable as the lot of the Oriyā had been under the Mughals, it was worse under the Marāthās. The misrule of these marauders presents a dismal scene of extortion, desolation and rapine; their cavalry harried the country at stated periods each year; and to quote Stirling's account, "their administration was fatal to the welfare of the people and the prosperity of the country, and exhibits a picture of misrule, anarchy, weakness, rapacity and violence combined, which makes one wonder how society can have kept together under so calamitous a tyranny." Fixed property did not exist; the people fled to the swamps and forests as an asylum from the Marāthā spearmen; and the land remaining untilled, the horrors of famine were added to the general misery. It would have been strange if the Oriyā character had not been affected by this grinding tyranny; and it is not surprising that the bitter experience of their forefathers should have discouraged thrift, promoted improvidence and tended to make the people a feeble and timid race.

There is, however, another influence at work which accounts very largely for their want of spirit and enterprise. From time immemorial they have been a priest-ridden race, kept in subjection by the Brāhmans and subject to all the paralyzing influences of religious superstition and caste prejudice. Nowhere else do the ancient caste* rules exercise such an influence. Men following precisely the same occupation are sometimes separated by so vast a social gulf that the slightest bodily contact with each other brings pollution; and the highest cannot touch any article that the lower has handled until it undergoes purification. Not only had the Brāhman the monopoly of education, but no one outside the priestly caste might plant even a cocoanut-tree. These profitable trees were only planted by non-Brahmanical hands after the advent of the missionaries, and the native Christian who had been the first to break the immemorial custom was regarded for many years as a man lying under the wrath of the gods. Even in more recent times, a Government official, who imagined that he would increase the revenue by planting cocoanut-trees along the Māchgaon canal, found, when the time came to sell the fruit, that the Brāhmans had forbidden any Hindu to purchase the nuts and was at last driven to get the best price he could from the native Christians in Cuttack. An equally striking instance of the strength of caste prejudice is the existence of the caste called Chhatra-khai, which is made up of the people who lost their caste in 1866 for eating in relief-kitchens (*chhatra*). The caste is divided into an upper and a lower sub-caste—the former comprising Brāhmans, Karans, Khandaits and Gop-Goālās, the latter consisting of the castes ranking below these in the social scale. Members of each sub-caste marry within that group, irrespective of the caste to which they originally belonged; but no intermarriage is possible between members of the two sub-castes.

PRINCIPAL CASTES.

Khandaits.

The Khandaits are by far the largest caste in the district, numbering, according to the census returns of 1901, over 375,000, or more than a sixth of the entire population. Besides contributing the largest share to the district population, the Khandaits have strong claims to be regarded as the most interesting caste in the district; and Cuttack may be aptly termed the land of the Khandaits, just as Purī is the district of the Chāsa caste. There is some difference of opinion as to the origin of the word Khandait. The general view is that it means swordsman (from *khandā*, a sword), but another explanation which has been put forward, and with much plausibility, is that Orissa was formerly divided into

* See Orissa by Sir W. W. Hunter, Vol. ii, pp. 139—141.

khandas, or groups of villages corresponding to the *pargana* of Muhammadan times, and that there was over each a headman called *khandapati*, which was subsequently corrupted to Khandait. Whatever may be the etymology* of the name, it is admitted that the Khandaits are the descendants of the people who formed the peasant militia under the ancient Rājās of Orissa. The armies of these chieftains consisted of various castes and races, the upper ranks being officered by men of good Aryan descent, while the lower ranks were recruited from the low castes alike of the hills and plains. As members of the militia, the Khandaits had to serve as soldiers in time of war, and in return they were given lands to hold under a strictly military tenure. Their characteristic occupation and the consequent relation with land all tended to alienate them from the communities to which they had originally belonged, and eventually led, on the establishment of a well-defined caste system, to the formation of the Khandait caste.

The different variety which these people exhibit and their free intercourse with some other castes tend to show that the Khandaits cannot trace their descent from a single origin and that the caste is only a heterogeneous group, which is perhaps made up at the one end of Aryan immigrants and at the other of recruits from a number of indigenous non-Aryan tribes. They are divided into two sub-castes—(i) The Mahānaik Khandaits and (ii) the ordinary Khandaits. The latter, who occupy the position of ordinary cultivators, appear to correspond to the rank and file of the old feudal militia, while the former, who hold large *jāgīr* tenures, may represent the officers of that body; an almost impassable gulf seems to exist between these two sub-castes, and there is nothing common between the two, except the name itself. On the other hand, cases of intermarriage between the Khandaits and members of other castes of equal standing are not at all rare. Karans, a fairly high caste of Aryan descent, are often found marrying members of Mahānaik Khandait families, and intermarriage between the Chāsas, who, as stated below, have an admixture of aboriginal blood, and the ordinary Khandaits is quite a common occurrence. The characteristic occupation of the Khandaits no longer existing, free intercourse has now sprung up between them and the Chāsas, and there is hardly anything at present to distinguish members of the two castes.

* In the Bengal Census Report of 1901 (p. 389) it is stated that "it is a significant fact that one of the caste *Santaks*, or devices endorsed on documents, is a *kānda* or arrow." It is stated, however, by Babu Jamini Mohan Das that the only caste which is known to use the *kānda* as a *Santak* is the low unclean caste of the Kandrās. The Khandait *Santak* is a *katāri* or dagger.

Although the Khandaits nowhere number less than 10 per cent. of the population, the inequality of their distribution is very marked. Whereas in Cuttack thāna they contribute only a tenth to the population, in the thāna of Aul or Rājābāri there are, in every hundred, no less than 28 Khandaits; the other thānas where the Khandait population exceeds the district average considerably are Patāmundai, Kendrāpāra and Tirtol. The distribution of this caste appears to correspond with the limits of the territories of the ancient feudatory Rājās of Orissa. It was their peasant militia which, as already stated, ultimately became the Khandait caste, and it is not surprising therefore that the Khandaits are found in largest numbers in those parts of the district which at one time formed the fiefs of the feudatory chiefs or in the adjoining tracts. Aul, Patāmundai and Tirtol were for centuries under the influence of native chiefs, and the Rājās of Kujang, Kanikā and Aul long exercised considerable power in the south-eastern and north-eastern parts of the district. The part of the district where the population of this caste falls considerably below the average is that included in the Cuttack thāna and its immediate neighbourhood. This tract was the chief portion of the Mughalbandi, or crown lands, as opposed to the military fiefs of the hill-country to the west and of the river jungles and marshes to the east.

Chāsas. The Chāsas, with a strength of nearly 266,000 persons, are next to the Khandaits the most numerous caste in the district. They are known to be recruited mainly from various aboriginal tribes, and the process of accretion seems to be still going on. As their name implies, they are an agricultural caste, the members of which almost all hold land as occupancy ryots or work for others as field-labourers. Like the Khandaits, they are the well-to-do peasantry of the villages, and are among the most skilful of all the Oriyā cultivators. Their distribution appears to be governed by the hereditary occupation of the caste, as they congregate most thickly in the fertile deltaic plains and are least numerous on the west, where there is a rocky fringe of hills and sterile uplands, and on the extreme east, where there are marshy jungles quite unfit for the plough. The thānas of Cuttack, Sālīpur, Tirtol and Jagatsinghpur, lying between the hills on the west and the swamps on the east, are the parts best adapted for cultivation; and these naturally are the thānas in which the Chāsas are mostly to be found. They are divided into four sects, the Orh, Benātiyā, Chukuliyā and Sukuliyā, of which the Benātiyā stands first in rank, and the Sukuliyā and Chukuliyā lowest; there is no material difference between the latter, except that the

Sukuliyās do not allow their females to appear in public. In this district a Benātiyā may intermarry with the other sub-castes on payment of a fine to his *panchāyat*, and all the sub-castes may drink and smoke, but may not eat cooked rice together. The Orh Chāsas, it is said, were the first of the aboriginal tribes who settled in Orissa and began to cultivate the soil; and they allege that, as they were very numerous, the country was called after them. The Benātiyā is said to have been created from a tuft of *benā* grass, or to be descended from the early settlers who first made the land fit for cultivation by clearing the *benā* grass away.

The Brāhmans, though constituting by far the most powerful caste in Cuttack, are not nearly so numerous as the two castes mentioned above, as at the census of 1901 they amounted to only 194,692 persons. They belong to the Utkal class of Brāhmans, which is one of the five great territorial groups into which the Gaura Brāhmans of Northern India are divided. Antiquarian research has not yet been able to fix the time when this division took place, but it may perhaps be assumed that the colonies of Aryan Brāhmans were separated by local usage, and that this separation was marked by geographical limits before the wave of Buddhism passed over the Utkal country. Buddhism deprived the Brāhmans of their priestly functions and drove them to more worldly pursuits for their subsistence. Most of them resorted to agriculture, while a few are believed to have taken service as cooks in the temple of Jagannāth. In the 5th century A.D., the ruling dynasty revived the Brahmanical faith in Orissa, not by restoring the priestly functions to the degraded Brāhmans who, forsaking the Vedas, had turned cultivators and cooks, but by importing 10,000 Brāhmans of pure faith, fit to perform Vedic rites, from Kanauj, the greatest stronghold of Hinduism in Northern India. Tradition relates that these Brāhmans performed 10 great "horse-flesh sacrifices" (*Asvameda Jajna*) on the bank of the sacred Baitaranī near the town of Jājpur, and a flight of steps, called Dasāsvameda Ghāt, yet marks the spot near which the sacrifices were performed. These imported Brāhmans gradually spread over the whole of Orissa, and the colonies which they formed with the aid of royal grants of rent-free lands are still known as *Sāsans*. Brāhmans.

In course of time, however, the process which caused the original division of the Gaura Brāhmans into five groups was repeated, and two endogamous sub-divisions were formed on the two sides of the river Brāhmanī, the northern sub-division being called Jājpurotriya and the southern Dakshinotriya. Jājpur or Birajā Kshetra is the centre of the former, and still contains the largest proportion of Brāhmans in the district. Purī is the centre of the latter, though colonies of Dakshinotriya Brāhmans have crossed the boundary since the cleavage and settled in the northern region. Throughout Orissa, Brāhmans taboo wine, but those who worship the goddess Kālī are permitted to drink it, and the temple of the great goddess Birajā at Jājpur probably became a centre for the spread of this objectionable habit. This seems to be the only feasible explanation of the legend that the water of the sacred Baitaranī became wine and that the Jājpur Brāhmans degraded themselves by drinking it; and it is noticeable that the southern Brāhmans give this as a reason for considering the northern Brāhmans inferior to themselves.

It must have been increasingly difficult for a growing community to keep strictly within the limits of the religious duties prescribed by the Sāstras; and a further split was, therefore, caused between those in the enjoyment of royal patronage who continued to observe them, and those whom necessity forced to depart from them. Each territorial sub-division has thus been divided into two groups called Srotriya or Vaidik and Asrotriya or non-Vaidik. The former includes the Sāsani Brāhmans who depend, for their subsistence, chiefly on royal grants of rent-free lands, and the latter includes the following classes:—(1) Sāruā or Paniari, growers and sellers of vegetables; (2) Pāndā, Pūjāri or Deuliā, professional temple worshippers or cooks; and (3) Marhia, priests of low castes, who receive alms from the humble clients whom they serve and enjoy the notoriety of being fed first in all feasts connected with *prāyaschitta* or purification ceremonies. The Srotriyas do not intermarry with the Asrotriyas, and the latter have no intercourse with the degraded Māstāns or Mahāstāns of the pre-Buddhistic period. The non-Brahmanical occupations and titles of the latter mark them out as a class quite distinct from the rest of the Brāhmans of Orissa; they are called Balarāmgotrī, apparently from the fact that the plough is believed to be the distinctive weapon of the god Balarām.

The Utkal Brāhmans were originally all Sāktas, but now they all keep the *sālgrām* and worship the four gods Vishnu, Siva, Ganesh, and Sūrya, and the goddess Durgā. Chaitanya converted some of the Brāhmans to Vaishnavism, but even these converts worship the four gods and the goddess mentioned above on ceremonial occasions. The Grām Devatī receives the same degree of homage from this caste as she does from the other castes in Orissa. The ten *sanskāras* or purifying ceremonies are a distinctive feature in the life of the Utkal Brāhman. According to the Sāstras, they should be performed at different periods of life, but in Orissa all the ceremonies are performed at the time of *upanayana* or assumption of the sacred thread. The Utkal Brāhmans observe most rigidly the limits of age laid down in the Sāstras for the marriage of girls, giving them in marriage usually before ten and seldom after twelve, unlike the other high castes, the Kshattriyas, Karans and Khandaits, whose daughters are rarely married before twelve and are sometimes kept unmarried up to what is regarded as an advanced age even among educated reformers.

The Utkal Brāhmans have *gotras* indicative of descent from old *rishis*, like the other Brāhmans of Northern India. The *gotra* groups are all exogamous, and some of them have been further broken up by titles indicating descent from more recent ancestors. Below this again, there are still more sub-divisions leading to a system of hypergamy, which, however, is far less marked in Orissa than in Bengal. In this connection, mention may be made of the remarkable fact that among the Utkal Brāhmans traces are found of the existence of the totemistic beliefs common among the Dravidian races. A Brāhman of the Atreya *gotra* for instance will not sit on the skin of the deer or eat its flesh. A Brāhman of the Kaundinya *gotra* similarly does not sit on the skin of a tiger, and a Brāhman of the Gautama *gotra* offers special *pūjā* to the cow on the occasion of marriage. The usage is explained, not by any direct descent from the animals revered, but by the legend that the *gotra rishis* who were invited to the *jajna* of Daksha fled in the disguise of animals when the *jajna* was broken up by Siva. This is no doubt a fiction invented to explain an aboriginal belief, which the Brāhmans apparently imbibed from the Dravidians with whom they came in contact. There is, however, no evidence that there was any actual infusion of Dravidian blood among the pure Aryans who were imported from Kanauj.

The Gauras, who number 139,587, are the great pastoral caste Gauras. of Orissa, corresponding to the Goālās in Bengal and Bihār. They nearly all possess cattle and are chiefly engaged in breeding cows and in selling milk, curd and *ghī*; about 25 per cent. are also engaged in agriculture, and some serve as *mulyās* or hired agricultural labourers. They also work as domestic servants and very largely follow the profession of *pālki*-bearers. There are several sub-castes, of which the Mathurāpurī ranks highest; in Balasore its members do not carry the *pālki*, but in Cuttack all Gauras will do so. The Gopapurī sub-caste is noticeable for the fact that its female members are almost the only women in Orissa who do not wear nose ornaments—a circumstance which, they pretend, connects them with Krishna's mythical milkmaids. The young women of both sub-castes prepare the butter and *ghī* which the elder ones take round for sale with their milk. Field labour of all kinds is eschewed by the Gaura women. The sub-caste known as Magadha ranks last, and is probably a recent accretion from some aboriginal tribe.

Pāns. The only other caste numbering more than 100,000 is that of the Pāns (103,205). They seem to have belonged originally to the aboriginal tribes, and are consequently found in large numbers in the western thānas of the district, Sālīpur, Jājpur, and Dharmshāla, which are close to the Garjāt hills. The social status of the caste is very low; they eat pork and fowls, drink wine, and repudiate the Hindu restrictions upon food. Their original occupation is said to be weaving, but they now mostly work as day-labourers, drummers and cane-weavers, and many of them have taken to cultivation. Their professed religion is a sort of bastard Hinduism, which in Cuttack inclines to Vaishnavism, each group of Pāns having their Pān Vaishnava who officiates as their priest. The veneer of Hinduism, however, has only recently been laid on, and beneath it may be perceived plentiful traces of the primitive Animism common to all Dravidian tribes.

CHAPTER IV.

PUBLIC HEALTH.

A COMPARISON of vital statistics for any lengthy periods is rendered impossible by the changes in the system of registering births and deaths which have taken place from time to time. In 1869 the duty of reporting deaths was imposed on the village *chaukīdārs*, and in 1876 the system was extended to births; but the returns received were so incomplete that they were soon discontinued, and, except in towns, deaths alone were registered until 1892, when the collection of statistics of births as well as of deaths was ordered, and the system now in vogue was introduced. VITAL STATISTICS.

The returns submitted since that year up to the end of 1904 show that there has been a substantial increase in the population. During the nine years ending in 1900 the reported births exceeded the deaths by more than 108,000; and the excess would probably have been still greater had the figures been available for 1891, which was a particularly healthy year. Even, however, if we take the average of the next nine years to represent the number of births in 1891, the aggregate of births from 1891 to 1900 would be 726,060, or 37·47 per cent. of the population of 1891. The number of deaths in the same period was 591,000, which gives a percentage of 30·50 on the population of 1891; and the excess of births over deaths was thus 6·97 per cent.,—a result which closely approximates to that obtained at the census. The corresponding percentage for the whole Province of Bengal (4·44) was considerably lower. During these ten years the death-rate only twice exceeded the birth-rate, once in 1892 and again in 1894. In the first year, when the death-rate was over 38 per mille, epidemics of cholera caused an unusually heavy mortality (10·98 per mille), and in 1894 the high death-rate of 35·13 per mille was due, in a great measure, to the bad crops in the preceding year and to the high price of food-grains. The people being scantily nourished, their weakened constitutions were less capable of withstanding the inroads of disease, while their distress was aggravated by the high floods of July, which forced the poorer classes in the affected tracts to supplement their scanty meals of rice by jungle products. In these circumstances, it is perhaps not surprising that the death-rate due to fever was 19·3 per mille, the highest recorded in this district. After this year there was a steady decrease in the death-rate till 1897, when there was again severe scarcity. The distress which the people suffered in consequence of the failure of the crops is clearly reflected both in the mortuary returns and in the birth-rate. The mortality at once rose from 27·16 to 33·39 per mille in 1898, and then in the following year dropped suddenly to the lowest ratio recorded (23·75); while the birth-rate fell from 43·21 to 36·74 in 1898, after which it rose to the highest figure yet reached, the number of births reported in 1899 representing a ratio of nearly 50 per 1,000.

The vital statistics since 1900 show that the health of the people has been good, except in 1901, when there were outbreaks of cholera and small-pox, which caused a large mortality and resulted in a slight excess in the number of deaths over the births. Since that year the birth-rate has steadily mounted, reaching the high figure of 43·28 per thousand of the population in 1904, while on the other hand the death-rate has fallen to 28·30,—a ratio considerably below the Provincial average of 32·45. The railway has contributed largely to this satisfactory result, one of the greatest immediate benefits it has conferred being the comparative immunity which the district has secured from the dissemination of epidemic disease by the crowds of pilgrims travelling to and from the temple of Jagannāth at Purī, who now go by rail instead of on foot.

PRINCIPAL DISEASES.

Fever.

According to the returns, by far the greatest number of deaths are due to fever, about one-half of the total mortality being ascribed to this cause. Here, however, as elsewhere, this is due mainly to the difficulty of diagnosing all but a few well-defined diseases. The village *chaukīdār*, who is responsible for the returns, is far from being a medical expert; he knows cholera, dysentery and small-pox, but most other complaints he classes indiscriminately as fever. It is impossible therefore to say what proportion of the total is attributable to malarial affections, though generally it may safely be assumed that wherever the mortality entered under this head is unusually high, the greater part of the excess over the normal is due to their prevalence. According to this criterion, malarial fever is not very common in Cuttack, as the proportion of the deaths shown as due to fever is relatively small in comparison with other parts of Bengal. During all the years for which mortuary returns are available, the death-rate has never reached 20 per mille; the average for the 5 years ending in 1903 was only 14·52 as compared with the mean ratio of 22·32 for the whole of Bengal; and in 1904 it was as low as 13·50, or nearly 9 per mille less than the Provincial average. Fever is most prevalent after the close of the rains when the water is gradually draining off the fields, but it visits almost every homestead at one time or other during the year; and even in such a healthy year as 1904 it is reported that no villages were immune throughout the year, though a few may have remained so for a few months at a time.

Cholera. The district has always been exceptionally liable to outbreaks of cholera, particularly before the opening of the railway, when it was either introduced or spread in the district by the pilgrims bound to and from the temple of Jagannāth. Writing in 1875 Sir W. W. Hunter says:—"The arrival of the pilgrim stream is, year after year, the signal for the ordinary sporadic cases to assume the dimensions of an epidemic. Cuttack, the capital of Orissa, suffered so regularly and so severely from the passage of the pilgrim army that the doctors having tried everything else, at last determined to shut the devotees entirely out of the city. The result upon the public health has been marvellous. Police are stationed at the entrance to the town, and warn the pilgrims that they must skirt round the municipal boundaries. A sanitary cordon is thus maintained, and Cuttack is now free from the annual calamity to which it was for centuries subject."

History of epidemics. Whatever may have been the effect of these precautionary measures in protecting the town of Cuttack, the district generally continued to suffer year after year from cholera, and the outbreaks were as frequent, widespread and severe as before. The decade ending in 1891 was particularly bad in this respect. In 1882 it was reported that the virulence and fatality of the disease was greater in Orissa than in any other part of Bengal, and it was worse in Cuttack than elsewhere in Orissa. The next year was also a bad cholera year, especially in the littoral thānas. In 1885 the mortality from cholera rose in the Dharmshāla thāna to 10·94, and in Aul to 10·71 per thousand, the disease having spread all over the district from the Patāmundai thāna, where the water-supply had been contaminated by the sea water brought in by the cyclone-wave, and by the numerous corpses of men and animals putrifying in it. Next year the death-rate again exceeded 8 per mille in these two thānas and in Aul; but the worst year in Cuttack, as in the rest of Orissa, was 1889, the cholera mortality ranging from 8·90 in Dharmshāla to 22·19 in Bānki.

Throughout the next 10 years cholera was seldom absent, but it assumed serious proportions only in 1892, when it accounted for 21,289 deaths, and in 1900, when the mortality was at the rate of 8·48 per 1,000. In 1892 the high death-rate was due to the long-continued drought which followed the cyclone of November 1891. The sea-wave which accompanied this cyclone rendered the water more or less brackish all along the sea-coast, and the drought aggravated this. The winter rice crop was moreover seriously damaged by the cyclone, and the *rabi* crop suffering from the subsequent drought, the people were obliged to take unwholesome food. Again in 1900 the evil effects of the early cessation of the rains of 1899, which were greatly intensified by the absence of spring rain in 1900, and the late commencement of the monsoon, brought about conditions favourable to cholera, and caused a mortality which was exceeded in only two other districts. Since that year there have been serious epidemics in 1901 and in 1903, the prevalence of the disease in these years being apparently due to the short rainfall which led to the pollution of the drinking water supply. In 1904 the death-rate from cholera was the lowest ever recorded (0·37 per mille).

Causes of its prevalence.

As regards the unusual prevalence of cholera in this district, reference has already been made to the pilgrims travelling to and from Purī as one of the main causes of the epidemics. After a long journey, during which they suffered many privations and hardships, the pilgrims arrived at their destination in an exhausted condition, only to suffer more hardships among the enormous crowds of people already assembled there. It will easily be imagined that the soil, atmosphere and water-supply soon became polluted, and that a combination of these circumstances told unfavourably on the health of the people, especially on those whose systems had already been lowered by a long and fatiguing journey and by deficient and bad food, and rendered them more liable to cholera. This they distributed in all directions on their return journey, principally by polluting with the germs of the disease the water-supply of the places at which they halted *en route*. In this way, cholera was introduced or spread in most of the localities which the pilgrims passed on their journey, and the disease finding a soil rendered congenial by a scanty rainfall and an impure water-supply, increased rapidly. On the other hand, the way in which cholera has almost invariably appeared as a consequence of deficient rainfall is very noticeable. In 1840 the rains ceased on the 24th September, and a terrible scourge of cholera followed which scarcely spared a single family. In 1883, 1885 and 1888 there was little or no rain in October or November, and the records of cholera shew that 7,482, 7,140 and 14,887 deaths occurred during the next seasons. The system of registration is no doubt more accurate than it formerly was, and this may account for the terrible figures of subsequent years, such as 1892, when there were 21,000 deaths; but the outbreaks were far too simultaneous and widespread and occurred at too great a distance from the pilgrim route to admit of the explanation that they were solely due to the passage of pilgrims. In more recent years, again, epidemics of cholera have been most severe when the district has suffered from late or short rainfall; and they are especially virulent when the winter crop is imperilled by drought in October, as in their eagerness to save the rice the ryots use every available drop of water for irrigation, with the result that by February there is little left for drinking purposes.

Small-pox has hitherto been almost an annual visitation, Small-pox. nearly as terrible in its effects as cholera; and Cuttack, like the other districts of Orissa, has long had an unenviable reputation for the frequency of its epidemics. Year after year the disease breaks out and causes a heavy mortality, though there is no doubt that a large number of the deaths are preventible. Inoculation is carried on through the district, and the disease is thus spread. When the outbreak of small-pox is mild in character, children of 5 to 10 years of age are wilfully exposed to the disease; and on returning to their own villages an outbreak results, which is not confined to those thus exposed, but spreads amongst the unprotected generally with disastrous results. A fuller account of the widespread practice of inoculation is given below, and it will suffice here to quote a description* given 30 years ago, which in some respects applies equally to the state of affairs existing at the present day. "Small-pox," it is said, "generally makes its appearance about the beginning of the year, and as a rule ends before the middle of April. The Civil Surgeon states that its regular appearance during these months is owing to the practice of inoculating with small-pox matter. The inoculators preserve the virus in cotton, and commence operations about the end of December or beginning of January. Small-pox thus spreads to the unprotected, and becomes general throughout the district. The Oriyās are perfectly regardless of contagion; and it is no uncommon sight to see people in the streets, or walking about the crowded market places, covered with the disease. Ancient prejudice stands in the way of vaccination, and even the more enlightened natives of Orissa will seldom allow their children to be touched with vaccine matter." The opposition to vaccination and the popularity of inoculation are still extremely great; and in

* Statistical Account of Bengal, Vol. XVIII, p. 235.

these circumstances it is not altogether surprising that in the three years ending in 1901, the average annual death-rate from small-pox was over 3 per mille. Since that year, however, the deaths from small-pox have largely decreased, and were only 701 in 1903 and 289 in 1904, as compared with 2,902 in 1902 and 7,253 in 1901; this result being attributed to the action taken against professional inoculators, of whom there were found to be 264 in the district.

Dysentery and diarrhœa. Dysentery and diarrhœa are unusually prevalent in Cuttack, and are responsible for a great number of deaths every year. In the 5 years ending in 1903, the mean ratio of mortality from this cause was 3·37 per thousand of the population, or more than five times as great as the Provincial average (·75); and though in 1904 the mortality was slightly less, only 3 other districts in Bengal returned a higher death-rate. The prevalence of this disease appears to be due to several general causes, such as neglect of the ordinary laws of hygiene, living in damp unventilated houses, exposure to the exhalations from decomposing organic or fœcal matter, chills and variations of climate, direct irritation of the intestines by indigestible food, and last, but not least, bad water and bad air.

Other diseases. There have been a few sporadic cases of plague since 1900, but the disease has never obtained any footing in the district, and up to the end of 1904 there have been only 109 deaths. Elephantiasis and hydrocele are extremely common. Statistics shewing what proportion of the population are affected are not available, but investigations made in 1902 among newly-admitted prisoners in the jail indicate how widespread these diseases are. Out of 1,194 prisoners examined 2·68 per cent. were found to be suffering from some form of elephantiasis, half of them having elephantiasis of the legs. Microscopical examination of the blood of 200 of the prisoners who were free from any form of elephantiasis shewed that filarias were present in 26 per cent. of the cases, and these filarias presented the character of the filaria nocturna. Altogether 270 of the 1,194 prisoners examined had hydrocele, and 25 out of each 100 suffering from hydrocele harboured the filaria nocturna in their systems. This number so closely approximates the 26 per cent. of the general body of the prisoners as to shew that, whatever might be the cause of the extraordinary prevalence of hydrocele, the filaria nocturna was not responsible.

Infirmities. In spite of the prevalence of small-pox, which often results in blindness, and in spite of the hot dry climate, glare and dust which are highly prejudicial to eyesight, Cuttack has a comparatively small blind population, the percentage of blind persons among males being 97 and among females 73 per 100,000. This percentage is, however, much greater than in either Balasore or Purī. Orissa stands high among the localities in which leprosy is prevalent, but the proportion of lepers in Cuttack (153 per 100,000 among males and 44 per 100,000 among females) is distinctly less than in the two adjoining districts. Insanity is relatively rare, as compared with Bengal proper, and the proportion of lunatics per 100,000 of each sex (excluding the inmates of the lunatic asylum) is only 24 males and 11 females.

SANITATION.

Outside the municipalities, sanitary efforts have until recently been almost a negligible quantity. Wells have been sunk and tanks cleaned, but there has been no serious attempt to improve the conditions prevailing in the mofussil villages. In the last few years, however, the District Board have been endeavouring to introduce a scheme of village sanitation by cleansing selected villages, in which the lanes are swept and the rubbish is removed. The Chairman, however, remarks:—"The usefulness of the scheme generally is limited by the amount of money that the Board can afford to allot. To effect any widespread improvement, the expenditure would have to be very considerable. The people are wedded to dirty ways, and expenditure on sanitation in a single year will not inculcate a spirit of cleanliness in a particular village. To keep a village clean and inculcate such a spirit, it will be necessary to continue spending money upon the same village or villages for several years, and thus a very long time must elapse before the advantages of sanitation can be taught and appreciated generally throughout the district." The fact is that the apathy of the people and the unwholesome habits to which they are rooted render the task of village sanitation on any appreciable scale most difficult. The houses throughout the district are built of mud dug up from the vicinity; and the result is that in the neighbourhood of almost every hut or house there is a dirty pit, filled to overflowing with water in the rainy season, and the receptacle of every description of filth. The consequent pollution of the water-supply and the effect on the general health of the villagers can be better imagined than described. In the towns the state of affairs is much better; a system of conservancy is kept up; night-soil and other refuse are removed; and steps are taken to protect the sources of water-supply. As a result of these measures, the health of the people in municipal areas, as shown by the vital statistics, is appreciably better than in the interior; in the year 1904 the town death-rate was more than 5 per mille, and in the previous 5 years nearly 3 per mille lower than that for the whole district. Even so, however, none of the towns possesses a pure and regular water-supply, and they all abound in filthy pits and hollows containing water of the foulest character and full of decaying vegetation which constitutes a standing menace to public health.

VACCINATION. Vaccination is unpopular among all classes in Orissa, where the people are more conservative, less enlightened and more wedded to superstitious beliefs than in the neighbouring Province of Bengal. Inoculation has, on the other hand, been practised for ages past, and the people believe in it. They see that its effects are serious, and they think that the powers of the goddess of small-pox are manifested by the eruption; while, as its substitute is not followed by an eruption or, as a rule, by fever, they distrust its powers of protection.

The practice of inoculation.* The profession of inoculator is hereditary among the Māstān Brāhmans, who also follow the calling of cultivators. They are found in scattered villages all over Orissa, and in this district there are several villages in the Sālīpur thāna entirely occupied by them. As inoculators the community and the villages in which they live are well known to the people generally, though the practice of inoculation is naturally most prevalent in the Sālīpur police circle. Their working season is usually a short one, extending from about the 1st November to the 1st March. Fees are paid according to the circumstances of the parents whose children are inoculated, and range from a minimum charge of 2 annas for a female and 4 annas for a male child to larger sums, in addition to which presents of cloth, rice, etc., are given. The income of an inoculator formerly is said to have varied from Rs. 100 to Rs. 300, but recently it has fallen and ranges from Rs. 25 to Rs. 200 per annum.

The material used is small-pox derived from a person recovering from an attack of variola discreta and removed on or about the 21st day of the disease: crusts of variola confluenta are not taken by skilful inoculators. After removal the crust is covered up with cotton wool and placed in a small hollow bamboo which is closed with a sola pith cork. When required for use—and this should be, if possible, within 3 or 4 days after removal—the cotton-wool containing the crust is moistened with water and squeezed on to a snail-shell; and the turbid fluid thus obtained is used for the operation. The instrument employed is a small piece of iron, shaped like a miniature country nail-parer, with a sharp edge; with this the skin is notched until blood just appears in the scratch, and the watery fluid mentioned above is

* The following account of inoculation is based on a report written by Major J. T. Calvert, I.M.S., Civil Surgeon, Cuttack.

then applied. Formerly male children were generally inoculated on the forearm, and female children on the upper arm; but the Pāns, the hereditary inoculators of some of the Tributary States, select a spot on the forehead between the eyebrows as the seat of inoculation. Recently, however, owing to the prohibition of the practice, it has been found necessary to select some less conspicuous place, such as the back part of the upper arm or knee, or the back of the hand.

Although there is no restriction regarding the age at which the operation may be performed, it usually takes place between the age of two and eight years, and in practice persons over 40 years of age are not subjected to it.

The religious aspect.

The operation is practically a religious ceremony. The day before it takes place a solemn offering is made to Sītalā, the goddess of small-pox, of which the essentials are cocoanut, milk, treacle, curd, cheese, plantains, turmeric, rice, *duba* grass, plum leaves and vermilion. This *pūjā* having been completed, the child is inoculated, and incantations are made to Sītalā until the scabs fall off. Four or five days after the operation the inoculator visits the child and takes his fees; and he comes again and offers *pūjā* to Sītalā, from the 9th to the 16th day, during the height of the eruption. Formerly this *pūjā* was performed openly with cornets and drums; but nowadays it takes place privately for fear of attracting attention.

After the operation the child is fed on cold rice and *feni* (a kind of sweetmeat), and has a bath daily until the eruption appears. The bath is then stopped, and rice, *dāl* and fried plantains form the dietary. During the period of convalescence the patient is humoured, dealt gently with, and never scolded, even if fractious, as it is believed that the deity presiding over small-pox is in the child's system, and any castigation or abuse might offend the goddess and draw down her wrath upon the child, in the form of confluent small-pox and death. It is also believed that the inoculators have the power of producing the exact number of eruptions which they promise before undertaking the operation; and they are credited with the power of allaying the intensity of the disease in a small-pox stricken patient. Their treatment consists in the administration of emetics and purgatives, by the action of which they believe the poison is washed away.

Its prevalence.

Inoculation is thus a regular profession, and the wide prevalence of the practice is sufficiently illustrated by the fact that out of 1,000 prisoners admitted into the Cuttack Jail in 1902, as many as 303 had been inoculated, 244 had had small-pox, 202 were unprotected, and only 251 had been vaccinated. It is difficult to put a stop to this mischievous practice, as the villagers are unwilling to come forward and give evidence against the inoculators. Not only, therefore, are convictions difficult to obtain, but until recently the punishments can hardly be said to have been of a deterrent nature, consisting as they did of fines which were easily paid. Of late years, however, a vigorous crusade has been carried on against this abuse, which has rapidly changed matters for the better. In spite of their pitiful protests that their means of livelihood would be taken from them, the inoculators were warned that their practice was prohibited by law and they must give it up, and that if they persisted in it, they would be liable to sentences of imprisonment. Those inoculators who disregarded this warning were prosecuted, and sentenced to various terms of imprisonment; the number of licensed vaccinators was also increased, in order to prevent the excuse that was formerly made that facilities for vaccination were not within the reach of all; notices were issued throughout the district pointing out that inoculation was prohibited; lists of unprotected children were prepared; and personal visits were paid by responsible officers to those localities in which opposition to vaccination was most intense. The result of these measures is seen in the increased number of vaccinations. Ten years ago the average number of successful vaccinations was under 16,000 annually, whereas in 1904 they were nearly 90,000, the average annual number of persons successfully vaccinated during the previous five years being 53,000, or the same as the total number in the four years 1892—96. The ratio of those successfully vaccinated, which averaged 26·74 per thousand of the population in the five years ending in 1903, is now 45·17, or nearly double that percentage; and instead of showing a ratio which was 4·71 lower than that for the whole Province, the district can now claim a percentage which is more than 9 per mille higher than the Provincial average.

MEDICAL INSTITUTIONS.

Thirty years ago there were only two institutions for affording charitable medical relief, besides the lunatic asylum, viz., the Cuttack dispensary or *annachhatra* hospital, and the Jājpur dispensary. The Cuttack dispensary was an institution connected with, or rather forming a part of, a general scheme for giving charitable aid to pilgrims and other poor people, and for supporting a number of *pāndās* or Hindu priests who keep up various temples and shrines in the neighbourhood of Cuttack. The *annachhatra* fund appears to have had its origin in assignments by the successive Hindu, Muhammadan, and Marāthā Governments for religious and charitable purposes; at the time of the first settlement of the district after its conquest, these charitable and religious assignments were continued as a charge on the revenues of the Province. Owing to the peculiar nature of this charity, less than half of the income was expended in the support of the dispensary proper, a fifth of the income was paid to the *pāndās* or in pensions, and half that amount in feeding the halt, lame, blind, lepers, etc., who assembled twice daily and received substantial meals each time. The female ward was generally filled with starving pilgrims or diseased prostitutes from the town, and the general ward was likewise full of pilgrims some of whom were half famished, while others were brought in the last stages of diarrhœa, dysentery and other wasting diseases. Naturally such an institution failed to attract respectable patients, when nearly all the indoor patients were pilgrims, or starving people picked up on the roads and brought in by the police, and the people of Cuttack, of the ordinary class of hospital patients, would hardly ever enter the hospital. The state of affairs at the Jājpur dispensary was very much the same; the patients were principally mendicants or starving pilgrims, and other classes looked upon the place as polluted and would never remain there.

Medical relief is now afforded at nc less than 12 dispensaries, as well as at the General Hospital at Cuttack; and the figures of attendance given in the Appendix will show how completely the old feeling against them has died away. Two of these dispensaries, the Cuttack Municipal Branch and Central Irrigation dispensaries, are in Cuttack town; two at the other sub-divisional head-quarters, Jājpur and Kendrāpāra, and the remainder at various places in the interior, viz., at Banki, Dharmshāla, Hukitālā, Jagatsinghpur, Nayā Bazar, Patāmundai, Ganjā and Rājnagar: the two last mentioned are maintained from the funds of the wards' estates at those places. These dispensaries have 34 beds, and in 1904 altogether 92,075 out-patients and 442 in-patients were treated, the number of operations performed being 4,516. A statement of the receipts and expenditure as well as of the principal diseases treated at each of these institutions in that year is given in the tables at the end of this Chapter.

Cuttack General Hospital.

The Cuttack General Hospital, which is the premier medical institution in Orissa, was established in 1874. It took the place of the old *annachhatra* hospital mentioned above, and a considerable portion of its income is derived from the *annachhatra* fund, a grant made by Government in lieu of the old endowment which it resumed; this fund brings in Rs. 3,700 a year after deducting the allowances of *pāndās* and paupers. It contains 58 beds for males and 22 beds for females; the

Lady Woodburn Hospital, which was opened in 1905 and which is intended to provide medical relief for respectable native women, forms part of it; and the accommodation for males will be increased when the Medical School has been removed to its new buildings. In 1904 the number of outdoor and indoor patients treated was 21,110 and 1,018 respectively, and 2,376 operations were performed. The great majority of the small minor operations are for the removal of elephantiasis of the scrotum or penis. The chief diseases treated in the out-patient department are diarrhœa, dysentery, malarial fevers, venereal diseases and eye diseases.

Cuttack Medical School.

The Cuttack Medical School was opened in 1876 with the object of providing Orissa with a supply of qualified native doctors. The course of study extends over 4 years and is uniform with that of the other Bengal Vernacular Medical Schools. The staff consists of the Superintendent, 3 Assistant Surgeons and 3 Hospital Assistants. During the last 5 years the number of students who have entered the school each year has varied from 46 to 67, while the number of those who have passed the qualifying examination at the end of the course has varied from 15 to 24 annually. The school was formerly housed in the General Hospital, but in 1904 a new class-room was built, and another building is now being erected, which will contain another class-room, a library and the museums and laboratories that are required by a modern medical school.

Cuttack Lunatic Asylum.

The Cuttack Lunatic Asylum was opened in 1864 and has accommodation for 43 male and 6 female lunatics. The average daily population since its opening has been 56, of whom 18 were criminal lunatics and 38 non-criminal lunatics, the average period of detention of the latter being one year. There is no separate accommodation for criminals and non-criminals; the asylum is not very favourably situated, being in the centre of the town and cramped for space; and it is proposed to abolish it and transfer its inmates to a central Provincial asylum.

Name of Dispensary.	Diseases treated.					Operations.
	Skin diseases.	Malarial fevers.	Intestinal worms.	Diseases of ear.	Venereal diseases.	
General Hospital, Cuttack	4,411	2,410	1,424	723	1,373	2,376
Municipal Dispensary, " ..	4,829	1,969	3,766	967	584	1,311
Central Irrigation Hospital, Cuttack	157	161	16	21	...	31
Bānki Dispensary ...	1,426	986	194	204	162	506
Dharmshāla " ...	1,234	624	567	285	29	579
Ganjā " ...	1,101	972	312	336	75	275
Hukitālā " ...	80	114	1	10	4	17
Jagatsinghpur " ...	2,892	484	48	89	76	222
Jājpur " ...	1,401	388	940	529	108	307
Kendrāpāra " ...	2,634	241	430	547	141	362
Nayā Bazar " ...	1,649	1,474	680	364	228	336
Patāmundai " ...	903	120	407	495	127	390
Rājnagar " ...	1,787	389	193	205	53	170
Total	24,514	10,332	8,978	4,775	2,960	6,882

Name of Dispensary.	Receipts.			Expenditure.	
	Government contributions	District and Municipal funds.	Subscriptions and other sources.	Establishment.	Medicines, diet, buildings, etc.
	Rs.	Rs.	Rs.	Rs.	Rs.
General Hospital, Cuttack	2,908	2,700	50,892	2,737	48,080
Municipal Dispensary, "	177	1,936	126	873	1,059
Central Irrigation Hospital, Cuttack ...	1,637	...	...	1,020	616
Bānki Dispensary	1,078	...	...	648	430
Dharmshāla " ...	154	851	290	1,013	282
Ganjā " ...	8	...	846	494	360
Hukitālā " ...	749	...	...	589	160
Jagatsinghpur " ...	254	285	118	401	256
Jājpur " ...	81	881	602	394	394
Kendrāpāra " ...	364	700	1,094	431	1,737
Nayā Bazar " ...	152	989	...	754	387
Patāmundai " ...	14	647	360	684	337
Rājnagar " ...	7	...	815	432	390
Total ...	7,583	8,989	55,143	10,470	54,488

CHAPTER V.

AGRICULTURE.

GENERAL CONDITIONS.

An account has been given in Chapter I of the three tracts into which the district is naturally divided, viz., the littoral, forming the sea face of the Bay of Bengal, the submontane, under the western hills, and between them a wide zone of highly fertile land intersected by a network of great rivers. To the east is a low-lying tract, which is of great natural fertility, where it is protected from the action of salt water; but a great part is impregnated with salt and unfit for cultivation, while much of the rest is exposed to damage from storm-waves. This belt of country contains treeless expanses of rice-fields and grass-lands, sloping down into a desolate jungly tract, full of swamps, saline creeks and impenetrable morasses, the haunt of wild hog and deer and of enormous crocodiles. To the west a large part of the surface consists of a series of low ranges, 10 to 15 miles in length, spreading out into infertile table-lands of ferruginous clay and laterite. It is a region of high sterile land and rocky hills, covered with bamboos and scrub jungle, and intersected by narrow though fertile valleys. Between these two tracts lie the wide alluvial plains forming the delta of the Mahānadī, Brāhmanī and Baitaranī rivers, where an extensive system of irrigation protects the crops from failure in seasons of drought and enables land to be cultivated that would otherwise remain barren. They present a gradual and steady slope from the high lands of the west to the sea, and a composition varying according to the relative proportion of the sand and silt of which they are formed. The surface is generally flat and presents the appearance of a dead level of rice-fields, but it is broken by the hills of Alti and Mātkatnagar in the centre, and is cut up by numerous river channels. In the west, where the mountains slope down to the plains, the lines of drainage are sufficiently marked by the great rivers, but in the delta proper the low levels lie not along the river courses, but in the valleys midway between them. The surface water gathers in many places in these intervening valleys into low marshes or temporary lakes, which are used during the dry season for the cultivation of the *dālua* or spring rice. In the central portion of this intermediate belt a large variety of crops are raised on the lands which are periodically enriched by river silt ; but along the western border and near the coast, winter rice is practically the only crop grown, as in the former tract the land is too high to receive deposits of silt, and in the latter tract the silt is deprived of most of its fertilizing power by the saline deposits of sea water.

Cuttack is primarily a land of abundant rainfall. Since 1860 the average registered fall for the year has been over 60 inches, it has occasionally been as great as 80 or 90 inches, and there have been only six occasions on which it was less than 50 inches. On the other hand, the rainfall is precarious, and an untimely or unequal distribution is liable to cause the partial or complete destruction of the crops, even if the actual fall does not fall short of the quantity required. A heavy shower in February or March is necessary to enable the land to be ploughed, but the most critical months are May, September and October. If the May showers, which are the precursors of the monsoon rains, do not fall, sowing may be prejudicially delayed; but deficiency in the rainfall in September and October is even more dangerous, as it affects the maturing of the staple rice crop. The most terrible famine the district has ever known was caused by the failure of the September and October rains in 1865 ; in 1896 with a rainfall very little below the normal, serious loss was caused by the cessation of the rains early in September ; and, on the other hand, the crops of 1876 and 1877 were saved by the rains in these months, in spite of the very scanty fall of 41·28 and 41·13 inches. On the whole, it may be said that a well-distributed rainfall of 40 inches is sufficient to secure the crop, provided that not less than 4 inches fall in October ; but in order to obtain a bumper crop at least 50 inches are required, of which 8 inches must fall in September and 6 inches in October. In the last 40 years, however, the fall of October has been less than 4 inches 15 times, and, generally speaking, the cultivators have to face the prospect of having once in every three or four years a rainfall less than the maximum compatible with the ripening of the crop, and of suffering a loss of a fourth to a half of the rice in the unirrigated lands. Besides this, the district is liable to inundation from the rivers overflowing their banks when swollen by heavy rainfall in the hills. It is only however when they are of an extraordinary height and of long duration, or when they occur so late as to render resowing impossible, that very serious and widespread damage is done by such floods. Provided that they are not too high or of long continuance, and that they come early in the

Rainfall.

season, they are productive of almost as much good as harm, as the fertilizing silt they leave behind renews the productive powers of the soil and assures excellent harvests.

Irrigation.

Owing to the ample supply of rainfall in ordinary years, irrigation is far less essential than in less favoured parts of the Province, and, except for the canals, it is little used. An account of the value of the canal system as affording protection against a failure or partial failure of the rains in years of drought has been given in Chapter VI; and it will suffice here to say that the area irrigated from this source is 170,000 acres and that the canal embankments protect about 500,000 acres. This area is practically all under rice, and water is taken from April to December, the demand for it being greatest in May and June, when it is required for ploughing the land, in July and August for loosening the soil at the roots of the young plants, and in October for the final ripening of the crop. Well water is used only for watering garden crops and betel plantations; and irrigation from streams and tanks is generally confined to the more valuable crops such as sugarcane, tobacco, and cotton; in most parts of the district it is only resorted to for paddy in October and November.

Water-lifts.

In the low-lying tracts near the sea water is taken from the small streams and creeks by means of the *tendā* or bamboo water-lift. This contrivance for raising water consists of two upright posts with a cross bar, which serves as a fulcrum on which a bamboo pole works; the latter is weighted at one end by a stone or mass of mud, and at the other a thin bamboo is fastened, with an earthen pot or bucket attached. When water is required the cultivator pulls down the bamboo pole till the bucket is immersed: as soon as the tension is relaxed, the weight attached to the lever raises the bucket of itself, and the water is then emptied into the *nodhā* or pipe, which is generally the hollowed trunk of a palm-tree, and is directed into the fields. When the field is any considerable height above the water, a platform is built on four stout bamboos on which a man stands to work the lever.

Where the water has only to be raised a few feet, it may be scooped up in a *senā*, a sort of basket made of split bamboo which two men use. Holding the ropes attached to either side, they swing it backwards, and bringing it down sharply into the water carry the forward motion of the swing through until the *senā*, now full of water, is raised to the level of the water-channel, when the contents are poured out. Another way of lifting water a short distance is with a scoop, called the *jantā*, which is made of a single piece of wood about 6 feet long, hollowed out and shaped like one-half of a canoe, the broad open end of which rests on the head of the waterchannel. The pointed closed end dips into the water, and when this is raised the water pours naturally into the channel. It may be worked by one man either directly or with the help of a bamboo crane and counterpoise, as is done with the *tendā*, but it cannot lift more than a couple of feet. It is not uncommon for two of these methods to be combined, the water being lifted by the *tendā* into a reservoir, and from that into the water-channel by a *senā* or *jantā*.

The arable land in the plains consists of alluvium in which sand and clay are intermixed in varying proportions; but the cultivators recognize a large number of different classes of soil, the names of which vary according to their situation, elevation and composition. In an ordinary village, the lands fall primarily under three main divisions according to their situation, viz., (1) The low lands retaining rain water and hence called *jala* or wet lands, on which winter rice is grown. Those lands predominate in the district and comprise about 70 per cent. of the whole cultivated area. (2) The high lands round the village homesteads, which, being enriched by manure and household refuse, have a blackish colour and are therefore called *kalā*; they are devoted to vegetables, cotton, jute, and other valuable crops. The homestead land is also known by the generic name of *gharbāri*, and the land lying between this and the fields is called *gāntalī*. (3) The river-side lands (*pāla*), which, being periodically fertilized by deposits of silt, are suitable for growing tobacco, cotton, mustard and other *rabi* crops. SOILS.*

Arable lands are also classified according to their elevation, the low-lying lands producing rice being called *khāl* and the high land *dhipa*; those situated on a level between the two are known as *madhiāna*. The low lands are further divided according to the difference in level into *jora*, *dera*, and *gahirā*. *Jora* is the name given to the hollows which collect the drainage of the surrounding high lands, and, being always waterlogged, are used only in the dry season or for very coarse varieties of rice; the *dera* lands are those situated at a higher level on the sides of the hollows; and the *gahirā* lands are those lying still further up. In hilly country the hollows lying between the upland jungle-covered ridges are known as *gorodā*. High lands which are not enriched by silt and cannot retain rain water are contemptuously referred to as waste land (*thenga*, *thengi* or *dānga*). The soils are again divided into

* This account is based on that given in the Report on the Agriculture of the District of Cuttack by N. N. Banerjee, M.R.A.C., to which I am also indebted for other information contained in this Chapter.

four great classes according to their composition, viz., (1) *Matāl* or clay lands, (2) *Dorasā* or loamy soils, (3) *Bāliā* or sandy lands, and (4) *Patu* or alluvial soils. The ryots, however, recognize a large number of minor distinctions and give different names to the soils according to the extent to which clay, sand, loam and silt predominate in their formation. (1) *Matāl* is the name given to all kinds of stiff clayey soils. Rice and sugarcane are the principal crops grown on them, but besides these, wheat, *bīrhi* and *kulthī* are also cultivated. (2) *Chikitā* is a strong sticky clay, which is almost too stiff to be used for successful cultivation, and grows for the most part coarse varieties of the *sārad* or winter rice. The outturn on such lands is said to be generally very poor. (3) *Chauria* is also a hard clay, which is very liable to cake on being exposed to the sun when ploughed. It generally cracks into hard blocks on getting dry, and is altogether an inferior soil. (4) *Dorasā* is a mixture of sand and clay in nearly equal parts. It is used for *biāli* or autumn rice and for all *rabi* crops. It is easily worked and is retentive of moisture. (5) *Telbāliā* is the name given to a loam which contains a larger admixture of sand than the *dorasā* lands. It is looser in texture, and being poorer, requires more manuring than the latter. (6) *Bāliāmatāl* is a loam with a large admixture of earth; in other words, a rich sandy loam. (7) *Rangāmāti* is a red ferruginous sandy loam occurring near laterite rocks. Like *telbāliā*, it requires a great deal of manuring before it can produce a good crop. (8) *Thenga jami* is an elevated sandy loam with very little moisture, which as a rule is allowed to lie waste, though sometimes ploughed up for growing *biāli* paddy, *māndiā*, and *kulthī*. The crops on such lands are necessarily very poor. (9) *Bāliā* is the name given to very loose sandy soils which grow the poorer kinds of *rabi* crops. (10) *Patu* is an alluvial soil, formed from silt deposited by floods. It is used for tobacco, jute, coriander and mustard, and is taken advantage of to grow all kinds of miscellaneous crops. (11) *Pankuā* is a black mud unmixed with sand, such as is found at the bottom of ponds and tanks. (12) *Rektimatāl* is a stiff rich soil. (13) *Pansiā* is a loose though comparatively fertile soil. (14) *Gengutiā*, as the name would signify, is applied to a clay containing an admixture of *genguti* or lime, which is met with near those river beds which contain nodulous limestone. An admixture of clay and limestone dust is in some places known as *khali*. (15) *Nunājami* is land which is more or less of a saline nature. It is generally of very little use; but a few varieties of *laghu sārad* rice are grown on it when it does not contain an excess of salt.

The staple crop is rice, which is grown on 1,195,000 acres,* or over 97 per cent. of the cultivated area. The varieties grown are very numerous, but they all fall under one of three heads according to the season at which they are sown and reaped, viz., (1) *Biāli* or early rice, sown in May or June and reaped in August and September; (2) *Sārad*, or winter rice, sown in June and July and harvested between October and January; and (3) *Dālua*, or spring rice, which is sown after the floods have subsided and is harvested in March and April. PRINCIPAL CROPS. Rice.

The most important of all these crops is the *sārad* or winter rice. It is divided according to the amount of water it requires into *guru* or heavy and *laghu* or light varieties, the *laghu* paddy being grown on moderately low lands which are wet or covered with six inches to a foot of water from June to October, while the *guru* paddy is grown at a lower level. The rice-fields vary in size ranging from small plots covering $\frac{1}{25}$th of an acre to large fields occupying an acre of ground. They are enclosed on all four sides by small ridges (*hira*) about a foot in height and breadth, in order that the rain water collected in these artificial shallows may keep the plants wet; otherwise the land losing its moisture, the plants would quickly wither and the crop be lost. Winter rice.

After the winter crop has been harvested in December, the cultivator is on the look out for the first shower of rain to plough his land. The time of ploughing necessarily depends on the rainfall, but if the cultivator is lucky, this occurs in February. As soon as the first shower falls, the country is covered with miserable-looking half-starved cattle dragging primitive ploughs, which as a rule never penetrate a foot below the surface of the soil. The land is ploughed as often as the weather and the resources of the cultivator permit, but as a rule four or five ploughings are considered sufficient. The soil after being turned up is exposed to the action of the sun and wind, and those lands which lie beyond the reach of the fertilizing river silt are manured. The peasant then waits for the showers which usher in the monsoon, and starts sowing as soon as they appear in May or June. The plants germinate in 15 days, and consequently the earlier the seed can be sown and the stronger the young plants are when the rains set in, the better is the chance of a good crop. During the latter half of June and the first half of July the growth of the rice is helped by the monsoon rains, and the cultivators have little to do but watch the young plants growing up, mend the small ridges round the fields, and do similar odd jobs. During the rest of Broadcast cultivation.

* The figures showing the area under various crops are the averages for the 5 years ending in 1903-04.

July and August when the plants have attained a height of about 15 inches, there is the important operation called *beusan* (literally changing of place) to be performed. This consists of driving the plough through the young rice in order to thoroughly loosen the soil at their roots; the rice plants are then firmly replanted by hand and a sort of blunt harrow is driven over the field to level and consolidate it. The ridges enclosing the fields are then finally strengthened, the grass cleared away from them, and the weeds removed. For these operations an ample supply of water is necessary, and if this is available and there is sufficient rainfall in September and October, a good harvest is secured in November and December.

From the preceding account it will be clear that the times of sowing and reploughing are two important periods when the *sārad* crop requires water, but by far the most critical period comes in the middle of October, when its fate depends entirely on there being enough water to mature it and to fill out the ear. At the first period no artificial irrigation is possible, and the people depend on rain water. At the second and third periods lands commanded by the canals, or about one-fourth of the cultivated area, can always get a plentiful supply of water, and under normal conditions the other lands also get sufficient rain water; but in years of deficient or unevenly distributed rainfall the people are obliged to irrigate the crop from every available natural or artificial reservoir. Fortunately, the people have now learnt to appreciate the value of canal water, but for many years they refused to realize that the loss of their crops from drought more than counterbalanced the saving of the water-rate. If the rain held off, the ryots grew daily more and more anxious; and though they began to discuss ways and means and the advisability of taking a lease of canal water, they took no steps in that direction; and it was not till the crop was irretrievably damaged to the extent of a fourth or a half that they took action. Then a panic would ensue, the whole country coming to the canal officer to demand water at a moment's notice over a vast area of parched and thirsty land, and, we are told, "ready to sign away their very lives in the urgency of the moment."

Trans-plantation.

Nearly all the *sārad* rice is broadcast, transplantation being an unpopular system of cultivation, as it involves more labour, and the transplanted seedlings are very delicate for the first month and liable to injury by flood and still more by drought; it is estimated that only 4 per cent. of the cultivated area grows transplanted rice. It is, however, admitted that, when successful, transplantation gives a larger yield; and it is resorted to for fields, especially for those under irrigation, which grow a *sārad* crop after *biāli*, to get rid of the wild paddy, called *bālungā*, to avoid the risk of early floods, and to replace the loss of the broadcast crop, if it is destroyed before the end of July. The seeds are sown either wet or dry in a nursery, which is generally a field near the village well manured and fenced in to keep off jackals and other animals. The land is carefully watered, and when the seedlings are a month old they are transplanted into the rice-field. The latter is prepared by ploughing and manuring in the same way as for broadcast rice, and is once again ploughed and harrowed before the young plants are planted. The seedlings are arranged in bunches of three or four plants with a small space between each bunch; the roots are carefully imbedded to the depth of a couple of inches; they are then left and require no further attention beyond a good weeding and a copious supply of water. The earlier the transplantation is done, the better the results, and the proper time is considered to be from the middle of June to the middle of July.

Biāli rice.

The *biāli* or early rice, which is always sown broadcast, ranks next to the *sārad* rice in importance. There are two main classes of *biāli*, viz., the early variety, called *sāthikā* from the fact that it comes to maturity 60 days from the date of sowing, and the *bara dhān*, ripening about a month later, which supplies the people with a food-grain only second in importance to the lowland paddy. Both varieties are grown on the higher lands of the villages and for preference in a light loamy soil; they are sown and reaped in the rainy season. The whole crop is more precarious than the winter rice, being injuriously affected by drought in June and July, and being also liable to destruction by heavy floods early in the season. A failure of this crop does not however affect the people very seriously, as loss can generally be recouped by a good harvest of winter rice. If the *biāli* is damaged by a deficiency of rain or by inundation, and there is no time for resowing, the lower lands at least can be sown with *sārad*, which with seasonable rainfall gives a good harvest, and so makes up for the loss occasioned by the failure of the early rice. On the other hand, if the rain is well distributed in the early part of the season but fails at its close, a bumper crop of *biāli* will in part at least compensate for the *sārad* crop being spoilt. *Biāli* rice is followed on high lands by pulses, generally *kulthī* or *birhi*, and on alluvial or homestead lands by mustard and linseed; in rich soil under irrigation or in favourable seasons *laghu* paddy is transplanted to the lands from which the *biāli* has been cut.

Dālua rice.

Dālua is a coarse variety of rice, which is grown in the Kharsuā and Brāhmanī estuaries on low swampy grounds and on lands too heavily waterlogged to yield *sārad*. Clay lands subject to tidal inundation are commonly chosen for the purpose, as irrigation is easy and the crop is not affected by saline matter. It is sown in winter and reaped in the spring, and therefore requires constant irrigation. Canal water is utilized for the purpose in the area commanded by the distributaries, but elsewhere it requires no artificial irrigation, as being planted along marshes and tidal waters a natural supply is generally available. The crop may be either transplanted or broadcast, but the former method is the more common. A nursery is selected in the corner of a field or tank, in which the seedlings remain till they are about a foot high; they are then imbedded in the rice-field which has been ploughed till it is a pulpy mass, and this is kept covered with water till the seed flowers. It ripens in March or April, and the crop is then cut. The area covered by it is not large, and it is only in the low-lying lands to the north of the district that it is a staple crop.

Outturn of rice.

From the experiments conducted at various times it appears that the outturn of clean rice per acre is 13½ maunds for irrigated and 12 maunds for unirrigated lands. The estimates of the Public Works Department show a much greater difference between the yield of irrigated and unirrigated lands, but their experiments are conducted in an area where all the best lands are irrigated and where the exclusion of river silt and systematic drainage have made a supply of water absolutely necessary. Outside this area the heaviest crops are raised on lands rich in river silt; and in ordinary years the average outturn of these lands is not very much less than in the irrigated country.

Other cereals and pulses.

After rice the most important cereal is *māndiā* (*Eleusine Coracana*) covering 15,100 acres, mostly in the central portion of the district. It is a cereal with a small reddish grain, which is grown on high light and inferior soils on which *biāli* rice would hardly succeed. Sometimes it is sown broadcast in May, but more often it is first sown in seed-beds and then transplanted in June. The crop, which requires good rain in June and July, ripens in August and September, and yields about 6 to 8 maunds of seed to the acre. This is ground into flour and eaten with cakes and rice by the lower classes; it is said to have the merit of producing such a feeling of satiety that after a full meal a man is not inclined to eat again for 24 hours.

Wheat is grown to a small extent on upland fields after rice, and generally on loamy or silt-covered soil; it occupies only 1,500 acres. It is sown broadcast in October or November and is reaped in April; as a rule it is left to grow up in the meantime as best it can. Barley is grown on light sandy lands, especially in areas exposed to inundation, and occupies the land from November to March. The area under this crop is insignificant, being only 1,000 acres. Other food-grains, including pulses, are produced on 114,600 acres; the only other cereal which need be mentioned is *chīna* (*Panicum miliaceum*), which is more largely grown than wheat or barley, but is of little importance; and after rice it is on the pulses sown in the autumn and harvested from January to April that the people depend. The commonest and least valued of these is the pulse called *kulthī* (*Dolichos biflorus*), grown on poor lands after *māndiā* and early rice, or on yet higher sandy lands which produce no other crop. The seed, which is like a dark flat pointed bean, makes an excellent food for cattle and horses; it is also boiled and eaten with rice by the poorer classes, and is even taken alone in times of scarcity. *Bīrhi* (*Phaseolus radiatus*) is a more valuable crop grown after *biāli* rice where the land is rich enough, and is found chiefly in inundated areas. It yields a little round pea, which is given to cattle and also eaten as a pulse in the form of *dāl*. *Muga* (*Phaseolus Mungo*) is the pulse most largely consumed by the better classes. Though not so general in its distribution as *bīrhi* or *kulthī*, it is the commonest *rabi* crop grown on *sārad* lands; it is found chiefly in the flooded tracts in the south of the district. The only other pulse calling for separate mention is *harar* (*Cajanus indicus*), of which there are two varieties—the *deo* or *nāli harar* grown on homestead lands and the *chaitā harar* raised on river-side lands after *biāli* paddy. The former variety is found in the west and centre of the district, and the latter is most largely grown on the border-land between Cuttack and Purī, and in the Brāhmanī basin, where it was the only crop which saved the people from starvation in many villages during the scarcity of 1896-97.

Oil-seeds.

Of all the oil-seeds, mustard and rape cover the largest area, being grown on 13,600 acres. Mustard is grown on rich loamy after a crop of rice and is commonest in the south of the district. It is one of the most valued of the *rabi* crops and its oil is used for anointing and cooking purposes. *Til* or gingelly is raised on 5,200 acres, linseed on 4,000 acres, and the total area under all other oil-seed crops is 13,000 acres; the most important of these is the castor-oil plant, which is usually found on homestead lands or in sandy fields along the beds of rivers, it being a peculiarity of the plant that it will grow in a depth of sand which would kill other crops.

Fibres.

The two chief fibre crops are cotton and jute. Cotton, which occupies 7,100 acres, is generally grown on homestead lands which can be watered from tanks and canals, or on rich alluvial soil by the riverside. The cultivation of jute is confined to the irrigated area in central Cuttack and Jājpur, and the area devoted to it is small (7,800 acres). Like cotton, it is grown on homestead or riverside land, where an extra rent has generally to be paid for the privilege of cultivating it.

Dyes.

Indigo is the only dye grown in the district. It was introduced by a Muhammadan merchant who established a factory at Kenduapatnā near the Kendrāpāra canal, but the area under it (200 acres) is insignificant.

Tobacco.

The tobacco plant, commonly called the *dhuānpatra* or smoke leaf, is one of the most valuable crops grown in the district, though the total area (7,500 acres) given up to it is small. It requires a rich loamy soil and a plentiful supply of water, and is only raised on rich silt-covered lands, on the banks of rivers, and in the depressions of the big flooded *pāts* or drainage lines. The finest leaves are obtained on a sandy sub-soil with a thin covering of pure silt. The profits of tobacco cultivation are very large, and it is estimated that one acre will bring in a net profit of Rs. 75. It is not possible however for one man to cultivate more than a quarter of an acre, owing to the unremitting toil it requires, and people having larger areas fit for tobacco sublet them to their neighbours. It is such a paying crop that many villages in southern Cuttack, especially in the *parganas* of Saibir, Deogaon and Sailo, depend on it for the payment of their rent.

Miscellaneous crops. Sugarcane.

Sugarcane, which covers 5,000 acres, requires a loamy soil and is grown generally on lands near the village and within easy reach of canal irrigation, or on the edges of natural water-courses, where the land is out of the range of canal water. It is a crop requiring incessant attention and involving a large expenditure of time, labour and money. The field has to be ploughed some thirty times and richly manured before the cuttings are planted in January or February. Then constant irrigation is necessary, and the soil has to be loosened and oil-cake and mustard oil applied to the roots. These processes are repeated at intervals, the land being irrigated so as to keep it continually moist; and after the fourth application of oil-cake in May or June, the soil is loosened by the plough and the land weeded. The stems are then wrapped in sugarcane leaves and tied up; after another weeding in August the leaves are bound together and the plants tied together in fours to give them greater power to resist the storms. Finally, in December the canes are cut down and the juice is extracted; the mills used for this purpose are extremely primitive, and the use of the improved Bihiā roller mill has not yet become general.

The cultivation of the climbing vine called *pān* (*Piper betel*), the leaves of which are used to wrap up the *supāri* or areca-nut chewed by natives of all ranks and classes, is not extensive, but its history is of some interest. It was introduced by some men of the Bārui caste who came from Bengal and settled down in Cuttack. It is still grown for the most part by men of this caste, but it is no longer confined to them, as the profits of the crop have attracted other castes, and now Khandaits, and indeed all castes but Brāhmans, cultivate it. The finest *pān* is grown at Barkud in the Kujang estate, where the immigrant Bāruis first settled; and the greater portion of the district, especially the markets in the Kendrāpāra sub-division, is supplied with betel-leaf from their plantations. There are also valuable gardens in Kodindā close to Cuttack and in the Jājpur sub-division, as well as in other parts of the district, but they are not so well known as the plantations of Barkud, which are jealously guarded from intruders, as the delicate plants, according to the growers, cannot bear any noise or disturbance. Under the sheds the leaves may be plucked and smelt, but not eaten, and women are not even allowed to bathe in the tanks and ponds from which the creepers are irrigated. The betel requires the most careful cultivation, but the crop is extremely valuable and the large profits amply repay the labour and expense which it entails; it is estimated that during the 18 years which may be taken as the average life of a garden—at the end of that time it grows to an unmanageable height and has to be abandoned—the cultivator obtains a net annual income of Rs. 131-10 for one *gunt* or ·08 acre of land. Betel.

The most important of all the garden crops is the *brinjal* or *baigun* (*Solanum melongena*), and its cultivation is very general. The *sāru* or caladium (*Colocasia antiquorum*) produces a tuber which is very largely eaten by the people. Onions are common, and cucumbers of many kinds are grown in homestead lands and may be seen climbing over the roofs of the houses in nearly every village. Pumpkins are also very generally grown; nearly every cultivator has a plant in his homestead, and they are also raised on a larger scale on sandy river-side lands. Potatoes have been recently introduced and are grown successfully at Cuttack town, where their cultivation is being quickly extended. The most popular fruit is the plantain, which is grown in nearly every part of the district; it is eaten as a fruit and also with curries, as, like the *brinjal*, Vegetables and fruits.

it forms the basis of most of the vegetable curries which please the palate of the Oriyā. Mangoes grow freely and form a very valuable addition to the food of the people during the hot weather, though their quality is decidedly inferior to the Mālda and Bombay varieties. Pine-apples are grown in many villages, but are not plentiful enough to form a very valuable article of food. Among other fruits are the *bel*, jack, tamarind, Indian plum, custard apple and papaya. Spices, turmeric, chillies, coriander and ginger are used largely in cooking, but the area they cover (1,200 acres) is too small to meet the demand, and there is consequently a considerable import, especially of the more valuable kinds. There are altogether 75,000 acres under garden crops and orchards.

EXTENSION OF CULTIVATION.

In the beginning of the 19th century the district had been reduced to a terrible state of desolation by the grinding tyranny of the Marāthās. The hereditary heads of the people had fled to the Garjāts where the independent tributary chiefs gave them protection in their hilly and jungly retreats; no land-holders could at first be found to engage for the lands; the ryots had found from bitter experience that they could get land on more favourable terms in the hills and had better prospects of enjoying the fruits of it; and the population was consequently insufficient to till the fields. A traveller who visited Cuttack in 1806 found himself in danger of wild beasts from the moment he entered the Province. Between Balasore and Cuttack, in a country now thickly populated and closely cultivated, he passed through a jungle abounding in tigers and required a guard of sepoys for the journey. Since that time cultivation has extended steadily under a settled government, though it was at first impeded by frequent droughts and by the injudicious settlements made in the early years of British administration; and now there is little land left for reclamation in the central portion of the district, where the pressure on the soil has almost reached its limit. The statistics available for the temporarily-settled estates afford a valuable index of the development of cultivation since the great settlement of 1837 was concluded. The cultivated area in these estates was then 697,000 acres, and the next 60 years witnessed an increase of 32 per cent., the area shown as under cultivation at the last settlement being 920,000 acres. The increase must have been still greater in the permanently-settled estates which lie on the seaboard and the hilly border region to the west; it has been most rapid in the north-east of Cuttack, but it has been general throughout the district and has steadily gone on with the growth of population, though it has occasionally been retarded by calamities such as the cyclones of 1885 and 1891. Many villages were deserted after the first disaster; the breaches it caused in the natural sea embankment of dunes or sand-hills were made worse by the cyclone of 1891; and areas which used to be cultivated were consequently swept by the salt water at high tides or when the tide was backed by a stiff breeze. The canal system does not appear to have been a special cause in the extension of cultivation; the increase has been no greater in the protected and irrigated areas than elsewhere; and the enquiries made on the subject have failed to elicit any evidence of a substantial extension of cultivation to lands which but for the canal water were not likely to have been reclaimed. At the present day the normal area* under cultivation in the whole district is 1,223,500 acres, of which 273,300 acres are twice cropped. Altogether 863,770 acres are not available for cultivation, but there are still 155,000 acres of culturable waste; much of this lies in the south-west of the district where the soil is very poor, or consists of scrub jungle at the foot of the hills; and it is doubtful if it will be brought under the plough for many years to come.

Improved methods of cultivation.

The Oriyā is a very conservative cultivator and has an apathetic indifference to agricultural improvements. Various experiments have been made from time to time in the Government and Wards' estates with new crops and modern implements, but these experiments have had little effect on cultivation generally. Various new crops have been tried such as potatoes, ground-nut, Nankin cotton, Buxar wheat, long-stemmed rice, sugarcane, etc., but, with the exception of sugarcane and potatoes, they have not made much way with the ryots. Efforts have also been made to introduce improved implements like Sibpur soil-inverting ploughs and the Bihiā sugarcane crushing mills, but with little success. The people are still wedded to the heavy old-fashioned Cuttack ploughs with two sides shaped like mould-boards which give them the appearance of ridging ploughs; and nothing shows their conservatism more clearly than their failure to adopt the improved sugarcane mills which have become popular almost everywhere else in Bengal. An experimental farm has recently (1904) been started at Cuttack, in order to carry out the recommendations of the Indian Irrigation Commission and to show the cultivators what can be done with water always at command; but it is too early as yet to say what success is likely to attend this venture. No advances have been made under the Land Improvement Loans Act, and little advantage has been taken of the Agriculturists' Loans Act; even in the lean years 1896—98 only Rs. 38,775 was advanced under the provisions of the latter Act.

* The average of the 5 years ending in 1093-04 is taken as the normal area.

Manures. The same conservatism is noticeable in the use of manure, as though the Oriyā is to a certain extent alive to its advantages, he will not use it unless his ancestors have done so, and applies it less freely than the cultivators in other districts. As in other parts of Bengal, cow-dung is the most important manure, but its value is much diminished by the negligent manner in which it is stored, and the feeding of cattle is so poor that it is not rich in manurial constituents. Besides this, a great deal is lost by its conversion into fuel cakes, as except in a few favoured localities firewood is scarce and its high price renders its use prohibitive for the ryots. For the most part, therefore, cow-dung only finds its way to the soil in the form of ashes; and the only other manure in common use consists of household refuse. These manures are spread on the rice lands at the time of the first ploughing, and are also applied to sugarcane, betel and vegetables. Oil-cake is also occasionally used as a top-dressing for these valuable crops. A strong prejudice exists against the use of night-soil and bone-meal, and chemical manures are practically unknown. The feeling against the use of bone-meal is particularly intense. In selecting a site for a building the greatest care is taken to remove all bones that the land may contain, as they are supposed to bring about ill-fortune and to cause the inmates of the house to die without heirs. The more superstitious even go through certain ablutions and ceremonies before re-entering their houses, if they happen to stumble across a bone in their fields.

Rotation. The scientific rotation of crops is not adopted as a principle of cultivation, but as a matter of practice rotation is observed in the case of the more exhausting crops. Sugarcane is never grown on the same land year after year, and when cultivated on *sārad* rice lands, it is alternated with paddy or follows a fallow, and is only grown on the same land once in four years. The lighter soils which bear early rice usually yield two harvests. When the crop has been harvested, the land is prepared for the usual *rabi* crops of pulse, wheat, barley, etc., and it is a common practice to grow *bīrhi* after the *biāli* crop and then to use the land for sugarcane. The mixture of pulses and cereals serves the purpose of rotation, as the pulses belong to the leguminous family and enrich the soil with nitrogen.

CATTLE. The cattle are similar to those found in the southern districts of Lower Bengal, but, owing to deficiency of pasture, the stock is generally poor. Some improvement has however been effected in the towns and a few places in the rural areas by crossing the local breeds with bulls imported from up-country. Pasture grounds abound on the seaboard and along the foot of the hills. During the hot weather large herds of cattle are grazed in the low-lying lands of Kujang, Kanikā and other estates on the coast, and are driven up to the jungly uplands on the west in the rains. Elsewhere the ground retains little moisture during the hot weather, and the grass being parched up by the burning sun, fodder is scarce. Cultivation has encroached on the grazing lands for many years past, though much has been done in the course of the recent settlement to reserve lands for pasturage; and the cattle have to be content with the dry stubble of the fields and such scanty herbage as the road-sides, river-banks, tank-banks and the boundary ridges of the fields afford. A cheap and abundant supply of *bīrhi* and *kulthī* is always available, but though these pulses make an excellent food for cattle, very few can afford to give them; while even the straw which might eke out the scanty supply of grass is largely used for thatching purposes. In the dry months therefore the cattle have only what they can pick up in the fields, though they are partly stall-fed on chopped rice straw while at work; they are generally underfed and miserably housed, and no attempt is made to improve the breed.

The sheep bred in the district are small in size with a short rough wool. Goats abound but are also small, though the breed imported from the south are somewhat larger. Pigs of the omnivorous kind found everywhere in Bengal are bred by Ghusuriās. The only horses are the usual indigenous ponies; they are few in number, undersized and incapable of much hard work.

The diseases most prevalent among cattle are rinderpest and foot-and-mouth disease. In 1903-04, 1,042 cases of foot-and-mouth disease were treated by itinerant Veterinary Assistants, and there were altogether 1,241 cases of rinderpest—a total exceeded only in Khulnā and Palāmau; but in 1904-05 these cases fell to 384. Veterinary assistance is also afforded at a dispensary in Cuttack town, where out-patients are treated; 88 horses and ponies and 481 cattle were under treatment in 1904-05.

CHAPTER VI.

EMBANKMENTS AND CANALS.

NECESSITY OF PROTECTIVE WORKS.

OWING to a well-known peculiarity of alluvial rivers, the water which is poured down upon the plains from the western hills greatly exceeds the volume which the lower channels are able to carry off. The rivers issue from the hills heavily laden with silt, which they deposit when their velocity is checked by the almost dead level of the delta; the fall in the Mahānadī and Baitaranī averaging from about 2 feet per mile where they enter the plains to 9 inches at tidal water, while that of the Brāhmanī is still less and does not anywhere exceed 14 inches per mile. The same process is repeated in the numerous channels into which they divide before they reach the sea; and their beds thus becoming gradually shallower, their capacity of discharge is greatly reduced. The most noticeable feature of this portion of their course is the meagre stream of water they bring down in the dry season, as compared with their great breadth, the shallowness of their beds, and their paroxysmal violence at periods of flood. In the hot weather they are nearly dry, and their beds consist of vast level stretches of sand, striped by long reaches of land-locked water, through which small streams meander from bank to bank. But in the rainy season, and especially after a storm has burst in Central India, they present an extraordinary contrast. These three great rivers collect the drainage of over 65,000 square miles; the entire rainfall of this enormous catchment area requires to find an outlet towards the sea; and the rivers rising with great rapidity dash down their concentrated floods on the small deltaic area of 3,600 square miles. The level strip between the mountains and the sea, which in itself has a rainfall of 60 inches in the year, has therefore to find an exit for the drainage of a territory of 15 times its own area; and the distributaries and channels often prove insufficient to carry off this enormous volume of water.

In their upper reaches they have a rapid flow and carry away the soil, but when they reach the level plains, their speed is reduced, and their torpid current is no longer able to support the solid matter hitherto held in suspension. They accordingly deposit it in their beds and on their banks, which are in this way raised above the level of the surrounding country. In fact, they practically run on ridges; and as their lower reaches have not a sufficient capacity for the vast amount of water they bring down in flood, they spill over their banks to a greater or less degree according to the chances of the season. In very high floods the excess discharge would inundate the surrounding country, were it not for the embankments; but even with the embankments a very large proportion pours down upon the rice-fields. Fortunately, the periods during which these vast rivers remain in high flood are not usually of long duration. The source of the Brāhmanī supply being more local, it both rises and falls more rapidly than the other two, the floods in it rarely lasting above 3 days; floods in the Baitaranī commonly last 3 or 5 days; and of the three rivers the Mahānadī takes longest to rise and remains longest in flood. In the rainy season therefore they would be liable to devastate the delta if left without control; and on the other hand, they fail to yield a trustworthy supply of water in the hot weather. The maximum recorded discharge of the Mahānadī is about 1,600,000 cubic feet per second, the average of the rainy season being about a third of this amount; in the Brāhmanī it is about 500,000 and in the Baitaranī 260,000 cubic feet per second; while the minimum discharge is 70, 129 and 204 cubic feet per second in the Baitaranī, Brāhmanī and Mahānadī respectively. An enormous mass of water, aggregating about 2,360,000 cubic feet per second, is thus thrown down in time of flood, while in the hot weather the total supply has been known to dwindle to 400 cubic feet per second; and the great problem which Government has to solve is how to prevent the rivers from destroying the crops during the rains, and how to husband them for agriculture and commerce during the dry season.

EMBANKMENTS.

The liability of the district to devastation by flood has been aptly described by Stirling. "The whole of the Mogulbandi," he says, "between the Chilka lake and the Brahmani river, is peculiarly subject to inundation from its proximity to the hills, the astonishing rapidity with which the torrents descend in the rains, and the strange conformation of the channels of some of the principal rivers, which are very broad within the hills, but divide soon after leaving them into a number of narrow streams. As an instance of rapid rise, it deserves to be recorded that, during the heavy rains of 1817, the waters of the Cajorī rose in one night a height of 18 feet, as ascertained by careful measurement. This immense volume of water, which was then perhaps one and-a-half mile in breadth by 30 or 40 feet depth, over-topped the general level of the town and station by a height of nearly 6 feet, and was only restrained from overwhelming them by a solid embankment faced with stone and supported by buttresses, the work of former Governments. The defence alluded to, however, called the revetment, has yielded in places within the memory of man, and the consequences were of course most tremendous. The Cuttack rivers are generally swollen to an extreme height about three times during each rainy season, and at such periods the crops and villages in many portions of the district are exposed to imminent hazard. To guard against the evil as much as practicable, embankments have been always maintained by Government, at a large expense. The embankments or *bunds* are solid mounds of earth, well sloped and turfed on either side, the principal ones measuring from 40 to 50 and 60 feet in breadth, and 8 to 16 in height. The havoc occasioned by the bursting of one of these large *bunds* is generally most serious. The torrent rushes through with a frightful roar and velocity, tearing up trees by the roots, prostrating houses, and washing clean away every trace of the labours of the peasantry. The devastations of the flood too are, in general, more permanently commemorated by a deposit of coarse sand, which renders the soil in the neighbourhood of the breach unfit for tillage for years afterwards."

Early history of the embankments.

Embankments intended to secure protection against destructive inundation appear to have existed in very early times, but whatever ancient works there were must have been isolated; and it has been held that they were rather of the nature of mounds on which villages were built, while the country generally was open to inundation. Under the Marāthā Government the zamīndārs were bound to maintain embankments, and for this purpose were allowed certain deductions from the revenue thay paid. This system, however, proved so eminently unsuccessful that from the earliest days of its administration the British Government undertook their maintenance and repair, and spent large sums on their upkeep. The old embankments were constructed at those places where the banks were specially low, in order to guard against the spill of the rivers during an ordinary flood. These embankments, by confining the spread of the water, raised its level, and so necessitated longer and stronger embankments to resist the floods; these new embankments in their turn again raised the level of the water, and thus led to the addition of more embankments, so that their construction was steadily progressive. In 1831 they came under the charge of the Public Works Department; and it is evident that there was then no regular system of protective works, nor does it appear that any attempt was made to systematize until 1855, or that anything was done beyond maintaining and improving the existing embankments. The matter was then forced upon the notice of Government; in the preceding three years very destructive floods occurred, which caused nearly 3,000 breaches; and finally the high flood of 1855, when the embankments were broken in 1,365 places, directed attention to the critical state of the revetment which protects the town of Cuttack, as well as to the broader question of destructive inundations in the district generally.

It was found that the head of the Kātjurī was enlarged greatly after each year's flood, and admitted a larger volume of water than its branches could possibly carry off, while at the same time the head of the Mahānadī was silting up and not carrying off its proper portion of the floods. In order to remedy this evil, a spur was constructed at Narāj with the object of regulating to some extent the relative discharges of the two rivers in accordance with the capacities of their channels: this spur was subsequently developed into the Narāj anicut, and as such still regulates the volume of flood entering the Kātjurī branch of the river. No systematic scheme was however sanctioned, as it was recognized that the proper control of such a vast river required the best engineering advice and experience; and it was decided not to enter on such a vast undertaking as the remodelling of the embankments till the whole question of the utilization of the water supply of the delta had been examined. Sir Arthur Cotton was accordingly deputed to report on the measures necessary for the management of the Mahānadī; and as a result of his investigations, he came to the conclusion that the small works already executed had commenced to have effect, and that it would be easy to turn back the larger portion of the water into the Mahānadī. He pointed out however that this expedient would only restore things to their former state, which was such that the whole delta was continually subject to awful droughts and flood; and he maintained that the only effectual remedy was to carry out a system of works which would completely regulate the waters of the Province, similar to those in the Godāvari and Kistna deltas.

Canal embankments. The canals which were eventually constructed were, accordingly, designed not only for irrigation and navigation, but also for protection from floods. The rivers run along the highest lines in the delta, so that when once they overflow their banks the surrounding country is inundated, and for a similar reason their margins present the most favourable alignments for canals intended for irrigation. Hence the general plan of the works was a series of canals leading off from the weirs at Cuttack, running along the margins of the great rivers, and having on the side next each river an embankment to keep out the floods. To control the flow of water down the respective canals and to regulate the discharge of the rivers in flood according to the natural capacities of the channels, extensive dams or anicuts of masonry were constructed with scouring sluices and sluices of discharge; and in order to afford relief from inundation, embanked escape channels were formed along the natural depressions which present themselves in the deltas of each river. This system was the first great attempt to grapple with the difficulty. There can be little doubt that, until the construction of the canals, the embankments were never thoroughly efficient; they were no doubt of some use in ordinary floods, and more or less protected villages from the strong currents; but in time of extraordinary floods they were of little use and were generally liable to be breached.

System of maintenance.

In the year 1866 there were about 510 miles of Government embankments and 248 miles of zamīndāri embankments in the district; but most of the latter, when originally constructed, were of insufficient height and strength to withstand heavy floods and had since fallen into disrepair and become useless. From 1866 onwards the embankments were much strengthened, but the question of the degree of efficiency in which they should be maintained was not raised till 1881. They had not been aligned on any scientific system, and it was physically impossible, without abandoning many of them and remodelling the remainder on an extensive scale, to render them capable of affording protection against high floods. A special enquiry was made regarding the expenditure required to put them in an efficient condition; and it was ascertained that in the case of the embankments on the Mahānadī alone the cost would be 45 lakhs of rupees, and that it would be necessary to construct embankments of such an enormous height that in practice it would not have been possible to hold them except at a very heavy cost. It was accordingly decided at the end of 1881 that the embankments should be kept up in the condition in which they then existed. Since that year the embankments have been maintained in much the same condition of efficiency; in repairing them care has been taken not to raise their height; and unauthorized additions have been prevented, as it was found that in previous years they had frequently been raised or lengthened, with the result that particular localities were protected, but that damage was caused elsewhere. A further examination of the embankments was therefore made in 1896 and 1897, in order that, when any obligations which might be held to rest on Government under the existing settlement might expire, only those embankments might be maintained which were productive of good or at least not harmful. Many embankments, it was found, were maintained simply because they were in charge of Government in 1881, and not because they were supposed to be of any real use to the country; in some cases there is no doubt that they were actually harmful, though they might afford some protection to particular places; and other embankments, though still nominally borne on the list, had already been practically abandoned, as the country they were supposed to protect was covered by the works constructed in connection with the canals. As a result of this examination, many embankments were abandoned; and Government now maintains under Act XXXII of 1855, 215 miles of embankments situated along the banks of the large rivers, which protect an area of about 1,000 square miles, while embankments extending over 265 miles are kept up in connection with the canals, which protect an additional area of 844 square miles. Vast sums have been expended from time to time on the maintenance of the embankments; from 1803 to 1830 over 8 lakhs was expended, of which half may be debited to Cuttack; from 1830 to 1866 over 7½ lakhs was spent in Cuttack alone, and in the next 30 years the expenditure amounted to 19 lakhs.

Value of the embankments.

How grave the danger of inundation used to be may be realized from the fact that in 1857 it was proposed to remove the cantonment, civil station and town of Cuttack to the left or north bank of the Mahānadī and to throw its site open to flood. Even at the present day it is impossible to assert that the embankments, as a whole, can withstand extraordinary floods or that the measures taken have been effectual in restoring the equilibrium of the river channels generally. On the other hand, there is no question that, in spite of these defects, they have proved of immense value to the district generally. Formerly the cottages used to remain under water for long periods during the rainy season, and the ryots had to remove themselves and all the moveable property they could take to the adjacent high lands or to the hills. There they had to wait patiently until the waters subsided, and then came down and repaired their houses. The canals have, to a great extent, put a stop to this, as their high embankments stand as a barrier to prevent the overflow of the water. Striking evidence of the protection now secured is afforded by the records of the high floods which have from time to time swept down on the district. The great flood of 1855 submerged nearly the whole country, though it did not rise beyond 123·48 on the Lālbāgh gauge, whereas at the present day a flood of this height need not be dreaded; and though the flood of 1866 caused a disastrous inundation, when the highest level reached on the Bellevue gauge was 125·50, a flood of similar height in 1895 caused very little damage.

CANALS. While embankments have existed from the earliest times in Cuttack, canals date from a comparatively late period in the history of the district. The first proposal to employ the rivers of Orissa for irrigation came from General Sir Arthur Cotton, who was deputed to visit the Province in 1858 with the object of giving advice as to the control of the flood waters of the Mahānadī. He recommended the construction of a complete system of irrigation and navigation canals, following the principles then being carried out in the deltas of the Godāvari and Kistna. He estimated that an area of $2\frac{1}{4}$ millions of acres might thus be irrigated, and that navigation might be opened up between Orissa, Midnapore and Calcutta, for a sum of 130 lakhs. Here, as elsewhere, Sir Arthur Cotton attached special importance to making the canals navigable, and pointed out how completely Orissa was cut off from the rest of India, destitute as it was of roads, railways or harbours, and traversed by a succession of formidable and unbridged rivers. In 1860 the East India Irrigation and Canal Company was formed for the purpose of carrying out the works in Orissa, and water was first supplied for irrigation in 1865. The works, however, were not sufficiently advanced to be of any real use in the terrible famine of 1866, though they supplied an excellent form of relief labour in the distressed districts. Before this it had become evident that the original estimate would be largely exceeded; and as the Company found it difficult to raise further funds, the Government of India purchased the whole of the works for the sum of 109 lakhs, and in 1869 the Company ceased to exist.

History of the canals.

From the first irrigation in Orissa made very slow progress. The works, however, proceeded, and in 1873 it was decided to provide for an irrigable area of 1,140,000 acres in Orissa, at an estimated cost of 441 lakhs. This area was to include 500,000 acres in the Balasore and Purī sections of the scheme, which had not then been put in hand and were soon after abandoned. The works sanctioned included the Tāldanda and Māchgaon canals for the irrigation of the lands between the Mahānadī and Kātjurī rivers; the Kendrāpāra and Patāmundai canals for the irrigation of the area between the Chitartala and the Birūpā; and three ranges of the High Level canal for the irrigation of the strip of country lying at the foot of the hills from Cuttack and Bhadrakh. By 1874 the greater part of this scheme was completed, but collections proved very disappointing, and in 1884 a revised scheme was approved for the extension of the Tāldanda and Mâchgaon canals and for the construction of new distributaries, bringing up the total estimate to Rs. 3,23,00,000, of which Rs. 2,02,00,000 had already been expended. The project then approved has been completed, and besides this the Māchgaon canal has been extended to the village of Nāgpur, one additional canal, with a total length of 7 miles, has been constructed from the junction of the Baitaranī and Burha to Jājpur, and a number of distributaries have been added. A channel, known as the Dudhai canal, taking off from the north end of the Brāhmanī weir, has also been recently constructed. It has been completed for 36 miles out of a sanctioned length of 46 miles, and is intended to irrigate about 12,000 acres of spring rice in the area between the Brāhmanī and Kharsuā rivers.

The general plan of the works is as follows. Near the point where each river bifurcates on debouching into the plains, a weir is constructed across the head of each branch, partly for the purpose of retaining the water at a suitable level for irrigation, and partly in order to distribute the flood discharge in suitable proportions between the different branches. From the flanks of these weirs marginal embankments run, if necessary, both up and down stream, so as to confine the floods to the river channels, and from the same points are led off the canals which conduct the water to the lands below. Provision is made by means of a network of smaller branch canals for the distribution of the water to the areas commanded. These channels, called distributaries, lead the water to within a certain distance of each village; and the more detailed distribution of the supply to the lands of each village is made by still smaller branches, termed village channels. The system now includes 315 miles of main canals and 1,167 miles of distributaries, including minor or village channels, of which all but 18 miles of canals and 65 miles of distributaries and minor channels lie in this district. The canals themselves branch off from Cuttack to the north, south and east, so that the four sides of the delta thus covered enclose a square, of which the northern boundary is the Baitaranī, the western the High Level canal, the southern the Māchgaon canal and the eastern the Bay of Bengal, while the Kendrāpāra canal may be said to form the diagonal of the square. Canal system.

The system derives its supply from seven great weirs with an aggregate length of 3½ miles, which, with the canal head sluices and entrance locks, constitute one of the most extensive systems of canal head works in India. Three of these weirs have been built in order to utilize and control the huge water-supply of the Mahānadī, viz., the Narāj, Mahānadī and Birūpā Weirs.

weirs; and the other four weirs are on the Brāhmanī and Baitaranī. The Narāj weir, which was constructed on the line of the old stone spur already mentioned, leaves the right bank of the Mahānadī below Narāj and runs obliquely down stream for a length of 3,833 feet until it meets the dividing embankment, which was originally intended to connect it with the island on which Cuttack stands. The Mahānadī weir runs across the head of the main branch of the Mahānadī at Jobra immediately below Cuttack, and supplies water to the Tāldanda canal and its branch, the Māchgaon canal. It has a length of 6,349 feet between its abutments and is pierced with two sets of scouring sluices, one of which has been placed at the south end of the weir, in order to prevent any accumulation of sand in front of the head sluices of the canal and the entrance to the Jobra lock, while the other is situated near the centre of the work, and serves the purpose of keeping a deep water-channel open for navigation in the pool above the weir. The Birūpā weir is situated on the river of that name about 1⅔ miles below its head; its length is 1,980 feet between abutments; and it is furnished with two sets of under-sluices, and supplies water to the Kendrāpāra canal system and the High Level canal, Range I. All these three weirs were constructed in a similar manner, and consist of a body wall of masonry, founded upon wells sunk into the sandy bed of the river, which is protected on the up-stream and down-stream sides by means of aprons of dry stone. The other four weirs are the Brāhmanī and Patiyā weirs on the Brāhmanī, and the Baitaranī and Burha weirs on the Baitaranī, the Baitaranī weir being in the Balasore district. They are intended to supply water to the second and third Ranges of the High Level canal, the Jājpur canal, and to the Dudhai canal on the left of the Brāhmanī; the Brāhmanī weir has a length of 4,000 feet and is situated at Jenāpur at the outfall of the first range, while the Patiyā weir, which has a length of only 783 feet, has been built on the Patiyā, immediately below the outfall of the Genda Nullah, at the head of the High Level canal, 2nd Range.

Main canals.

The canals which obtain a supply of water for irrigation and navigation from these seven weirs are—(1) The High Level canal, with one branch running to Jājpur and the other to Bhadrakh in the Balasore district. (2) The Kendrāpāra canal, with its extension to Jambu and two branches called the Gobri and Patāmundai canals; besides these, another canal, called the Gobri Extension canal, is supplied with water from the Kendrāpāra canal by means of the Patāmundai canal. (3) The Tāldanda canal, with its branch, the Māchgaon canal. Of the two main sections into which the district is divided by the rivers which traverse the delta, the tract between the main stream of the Mahānadī and the Brāhmanī is irrigated by the Patāmundai canal on the north, and the Kendrāpāra canal on the south, the Gobri canal forming a connecting link between them to the east. Both these systems draw their supply of water from the south flank of the anicut across the Birūpā, which also feeds the High Level canal. The anicut across the main branch of the Mahānadī feeds the Tāldanda and Māchgaon canals, which water the northern and southern edges of the tract between the Mahānadī and Kātjurī. All these canals maintain a high level along the banks of the rivers, which are always higher than the intermediate alluvial tracts.

The High Level canal. The High Level canal was designed to provide a navigable trade route between Cuttack and Calcutta, and also to irrigate the country through which it passes. It starts from above the left flank of the weir across the Birūpā, 1⅔ miles below the departure of that river from the main stream of the Mahānadī, and runs thence along the foot of the hills north-eastwards, through the Cuttack and Balasore districts. The original scheme was to carry the canal across the district of Midnapore to meet the Hooghly river at Ulubāria, below Calcutta, a total distance from the starting-point of 230 miles, so as to connect Cuttack with Calcutta by one long canal. This great scheme has however been abandoned, and only three ranges have been completed, of which the first and second, covering a total distance of 45½ miles, lie within this district, viz., Range I from the Birūpā to the Brāhmanī river, 33 miles long, and Range II from the Brāhmanī to the Baitaranī river, a distance of 12½ miles. The two ranges command an aggregate area of 57,495 acres, of which about half is actually irrigated. The High Level canal is the most picturesque of all the canals of Orissa, skirting the base of the wooded hills along the western boundary. The traveller looks eastward over almost boundless rice plains, the level surface of which is broken only by a few hills that here and there rise steeply from the surrounding country; while to the west is a vista of range upon range of rugged hill and valley in endless confusion.

The Jājpur canal, starting from the head-works at the point of bifurcation of the Baitaranī and Burha, runs 6½ miles to the town of Jājpur. It has a discharge of 600 cubic feet per second and it commands 70,000 acres. It is one of the youngest members of the Orissa system, and secures from drought the valuable rice-growing lands lying in the tract between the Baitaranī and Kharsuā, where formerly the low-lying villages could only be irrigated with brackish water from the creeks.

The Kendrāpāra canal.

The Kendrāpāra canal, which was opened in 1869, is the oldest and most important canal in the district. Taking off from the Birūpā river at Jagatpur just above the anicut, it skirts the northern bank of the Mahānadī and its tributary the Nūn, running nearly due east to Mārsāghāi, up to which point it is navigable. It has a total length of 39 miles and a discharge of 1,067 cubic feet per second. The area commanded by it is 106,159 acres, and its 23 distributaries are capable of watering 62,432 acres. It irrigates the country between the Mahānadī and the Gobri drainage channel, its right bank forming a protective embankment as well as a thoroughfare for the people. The country it commands comprises some of the most highly assessed *parganas* in the district. The Kendrāpāra Extension canal is a continuation of it, which runs between Mārsāghāi and the Jambu river, a distance of about 15 miles. It was originally constructed with the object of improving communications between Cuttack and False Point harbour, into which the Jambu flows, but it is also capable of supplying water for irrigation purposes to a small area.

The Gobri canal is a branch of the Kendrāpāra canal from which it takes off in the 28th mile. It has a total length of 15 miles and commands 18,850 acres, but the distributaries constructed can only irrigate 6,599 acres. It was originally intended to be a distributary of the latter canal, but was afterwards made navigable in order to facilitate communication between Cuttack and Chāndbāli, and it now forms part of the main route between the two places. The Gobri Extension canal is only 6 miles long, but commands an area of 12,717 acres, of which, however, only 5,174 acres can be irrigated by the distributaries constructed. It derives its water-supply from the Patāmundai canal, and forms the connecting link between the terminus of the Gobri canal on the Gandākiā river and the Brāhmanī at Alba.

The Patāmundai canal branches off from the Kendrāpāra canal just below the Birūpā head-works, and skirts the southern bank of that river and of the Brāhmanī river for a total length of 48 miles. It has a discharge of 885 cubic feet per second and commands an area of 54,800 acres, its distributaries being capable of irrigating 19,838 acres. It is provided only with weirs and is therefore impracticable for navigation; but it irrigates some of the richest lands in Orissa, and its left bank protects a large tract from the floods of the Birūpā and Brāhmanī rivers.

The Tāldanda canal.

The Tāldanda canal starts from the right bank of the Mahānadī immediately above the anicut at Jobra and runs in a south-eastern direction to Bīrbāti, where it gives off the Māchgaon branch Thence it runs along the southern bank of the Sukpaikā and the

Mahānadī for a total length of 52 miles. It has a discharge of 1,342 cubic feet per second, of which about half is taken off by the Māchgaon canal, and it commands 42,939 acres. It was designed for the purpose of irrigating the triangular tract of country between the Mahānadī and Kātjurī; but it is navigable by boats of a considerable size, and provides an alternative route from Cuttack to Chāndbāli *viâ* the Hansua creek.

The Māchgaon canal leaves the Tāldanda canal 7 miles south of Cuttack, and runs along the north bank of the Kātjurī and of its branch, the Alankā, for a distance of 32 miles; it has a discharge of 776 cubic feet per second and commands about 120,000 acres. It was originally intended to carry this canal as far as Māchgaon so as to run into the tidal water of the Devī river and thus establish connection with the sea, but this scheme was never carried out. It stops 6 miles short of Māchgaon, and there does not appear much probability that it will be extended to the termination at first proposed.

Canal revenue. When the project was first mooted, the most sanguine expectations were entertained as to the revenue the canals would yield; and in 1867 the Directors of the East India Irrigation and Canal Company estimated that the scheme would eventually return a net income equal to 21 per cent. of the outlay. These hopes soon proved delusive. It was found that the receipts did not cover the working expenses, while the interest on the loan steadily accumulated and quickly amounted to a sensible addition to the capital outlay. The hopes of a steadily increasing demand for water were dispelled, and though the outlay was very large, the incomings were insignificant. Six years after water was first offered to the people, irrigation was as far from general adoption as it was at first. By steadily refusing the water on the terms originally offered, the peasantry succeeded in beating down the rate, and the use of canal water then gradually extended. The works have, however, never been a success financially, and even up to the present day the outlay has proved unremunerative. The average annual working expenses for the five years ending in 1900-01 weɹe Rs. 4,91,830 and the average gross revenue from all sources wsa Rs. 4,67,913, so that the working expenses were not covered. In the last two years the position has improved a little, as the receipts exceeded the working expenses by Rs. 14,644 in 1902-03 and by Rs. 5,310 in 1903-04; but the total net revenue realized up to the end of the last year amounted to only 0·17 per cent. of the capital outlay, the total capital cost, exclusive of interest charges, was Rs. 2,65,62,647, and the interest alone amounted to Rs. 2,89,72,164.

Various causes have combined in falsifying the expectations which were originally entertained. The cost of the work was greater than was expected, the area under irrigation has not come up to the early forecasts, the navigation receipts have never been large, and the rates charged for irrigation are decidedly low. As already stated, Sir Arthur Cotton estimated that $2\frac{1}{4}$ millions of acres might be irrigated, whereas the average area is only about 200,000 acres. His estimate was doubtless a very rough one; and it probably included every acre of land within the Mahānadī delta and in the tract to be commanded by the Midnapore canal, without reference to the question whether it was cultivable or would take water. The areas shown in the revised project of 1873 again were estimated on the assumption, for which there was little warrant, that 500 acres in every square mile would be irrigated, so that the gross area commanded in Orissa would have been 1,140,000 acres, or 820,000 acres excluding the Purī and Balasore sections. The area actually commanded by the present system of distributaries is estimated at 528,534 acres, of which only 274,625 acres, or about 333 acres per square mile commanded, are irrigable; all of this is practically confined to this district, in which the culturable area commanded is 468,414 acres, of which 231,028 acres are irrigable. The balance is either uncultivable, or lies too low to take irrigation or too high to be irrigated otherwise than by lift, and lift irrigation is hardly practised at all. It is doubtful therefore whether the average area that can be irrigated by the present canals will ever exceed 250,000 acres, though there is no difficulty in regard to the water supply, which is generally sufficient for all the land which is ever likely to take irrigation.

Those who enthusiastically quoted the success of irrigation in the Madras deltas seem to have forgotten that in those tracts the rainfall does not exceed 40 inches, whereas in Orissa it amounts to 60 inches per annum. The normal rainfall being ample, the value of canal irrigation is exceptionally dependent on the character of the season; and the ryots do not consider canal irrigation so absolutely indispensable as to make it worth their while to pay anything but a small water-rate or to have all their land irrigated. Nothing shows more clearly why the canals have fallen so short of the expectations of revenue which were at first so general than the way in which the area under irrigation has varied and the people have refused to pay high water-rates. The question of the rates to be charged for water does not appear to have been considered at the time that the Company was formed; but after the works were taken over by Government in 1868, a scale of rates was notified which in practice proved to be prohibitive, viz., Rs. 6

per acre for sugarcane, Rs. 5 for certain other crops and Rs. 3 for any single crop not remaining more than six months in the ground. The people declined to pay such high rates, and were also afraid that irrigation would be made an excuse for enhancing the rents and revenue. To allay their apprehensions, a proclamation was issued by Government, declaring the water-rate to be wholly distinct from land revenue, and promising that at the next revision of the settlement no increased rate of assessment would be imposed on any lands by reason only of their being irrigated. These promises failed to produce much effect, and a much more effective inducement to take water was afforded by the gradual reduction of the rate to Re. 1-8 per acre. As a result of this measure, irrigation increased slowly, but on the whole steadily, and by 1876-77 there were 30,000 acres watered from the canals. There was then a great demand for the privilege of irrigation due to the high prices and scarcity of water in 1878, the year of the Madras famine, and to the introduction of the system of five year and one year leases. The area under irrigation rose suddenly to 98,000 acres, and increased to 133,000 acres in 1882-83. In 1883, however, there was a drop to 48,760 acres, as most of the leases expired and the people refused to renew them. The reason for this appears to be that in the eyes of the cultivator the chief value of the water lies not in any improvement it may render possible in the outturn of an ordinary year, but in the protection it affords in year of drought. This being so, the peasants are always disposed to put off renewing their engagements till a period of drought occurs; and this tendency was accentuated by the fact that in three out of the preceding five years the rainfall had been sufficient and timely, and consequently the benefit derived from the canal irrigation had been comparatively small. They soon began however to realize that the loss of their crops from drought more than counterbalanced the saving of the water-rate; the area slowly extended, and with the next quinquennial period a much larger number of leases were executed, and the maximum of 186,627 acres was reached in 1889-90. A few years of abundant and excessive rainfall brought the irrigated area down to 119,460 acres in 1896, when the drought created an universal demand for water, and the irrigated area rose at a bound to 207,015 acres. Since that year it has increased but little; the average area irrigated in the three years ending in March 1903 was 210,012, the maximum of 225,000 acres being reached in 1902-03; and in 1903-04 the area supplied was 210,161 acres.

Navigation. A very large sum was expended in making the canals first class navigation lines, as great importance was attached to this source of income. It was confidently expected that there would be a large traffic along these waterways, and that the receipts from tolls would give the State a profitable return for the money spent on them. It might, indeed, reasonably have been expected that, if navigation could be a success anywhere, it would be in Orissa. The country was in a terrible state of isolation without internal and external communications, and it was naturally anticipated that the canals would attract all the local trade and form cheap routes connecting all parts of the country. It is clear however that, after 30 years of a very fine system of navigable canals, the people have not taken to navigation. Carts and pack-bullocks still constitute the chief means of transport, and even when canals are available, the people seldom use boats. The navigation receipts have therefore always been an insignificant source of income, and since the opening of the railway they have still further diminished, falling from Rs. 1,94,100 in 1897-98 to Rs. 78,153 in 1901-02. The tollage rates have been recently reduced, and the tonnage borne along the canals has increased; but even so the tollage only amounted to Rs. 68,489 in 1903-04, and there does not appear to be any likelihood of navigation ever proving of much value as a source of income.

Canal administration. This large irrigation system is under the control of a Superintending Engineer, who is assisted by three Executive Engineers in charge of divisions. The latter are responsible for the maintenance of the canals and the conduct of irrigation operations; and a separate establishment is entertained for the collection of the revenue. For this purpose there is a revenue division in charge of a Special Deputy Collector, who sees to the assessment and collection of water-rates under the orders of the Superintending Engineer. The irrigated area is divided into blocks, the lease of all the lands in each block being arranged so as to lapse in the same year, while efforts are made in fixing the period of the lease of these blocks to see that leases for an equal area expire each year. Water is supplied to the cultivators on application on a prescribed form, the year being divided into three seasons, that is the hot weather, from March to June; *kharīf*, from the 16th June to the end of October, and *rabi*, from November to the end of March. Dates are fixed for each season, and a lease or permit granted for the season is only in force for that particular period. Besides these season leases, there are long-term leases, or leases for periods up to ten years, which are granted at a somewhat reduced rate, and secure a supply of water from the 16th June to the 31st March in each year. These long-term leases are only granted for compact blocks defined by well-marked boundaries of such a nature that the leased lands can be clearly distinguished from the adjoining unleased lands, and also so situated that unleased lands will not be ordinarily irrigated by water supplied for the land included in the block. These boundaries are mentioned in the application for the lease, on receipt of which a special report is submitted to the Executive Engineer. If the lease is approved, that officer issues orders for the block to be measured, and a detailed *khasrā*, or measurement of each cultivator's holding, is then made. The lease is finally approved by the Executive Engineer who issues the permit, but before this can be done, every cultivator who has fields within the block, must sign his name against the area which has been measured, and which will be assessed in his name. In order to admit of a block getting water for the first season, a provisional permit is granted for the season on the area originally applied for; this permit is cancelled when the long-lease permit is finally granted. Fields which cannot be ordinarily irrigated, or for which canal water is not ordinarily required, can be excluded from the block at the discretion of the Executive Engineer, such fields being duly noted in the *khasrā* or measurement paper. In these long-term leases water-rates are charged for the area measured and accepted by the cultivators, whether water is required or not. In *rabi* and hot-weather leases, water is supplied on application, and water-rates are levied on the actual areas irrigated, and not necessarily on those specified in the application. In order to assist the Canal Department as far as possible in the assessment and collection of water-rates, influential men of the village, called "representatives," are appointed on the approval of the majority of the cultivators concerned. Their duty is to assist in measurements, in procuring and attesting signatures to applications for leases, and in collecting the rates. In return for this work, they are entitled to free irrigation of the lands in their own occupation within the leased area up to a limit of 3 per cent. of the area assessed.

The present practice is to give long-term block leases, which often extend to 10 years, but to discriminate between the various classes of land forming a block. Thus lands lying so low that they never require irrigation, although water may often flow into them, are excluded from assessment; while a special rate of 8 annas per acre is charged on those lands which derive benefit from irrigation only in exceptionally dry years. The rate charged for other land, or the ruling rate, was formerly Re. 1-8 per acre, but it was raised to Re. 1-12 in 1902-03. Higher rates are charged for single season leases, or for water taken between 1st April and 16th June; but the average or all-round rate is about 3 annas less than the Water-rates.

ruling rate, and it amounted to Re. 1-5 only during the three years ending 1901-02. The long lease system is well suited to these canals, as it tends to prevent loss of revenue in seasons in which irrigation is not required, and it appears to be popular with the cultivators. The fact nevertheless remains that, after many years' nursing of the lease system, it has not been possible to induce the Oriya cultivator to pay an average rate of more than Re. 1-5 per acre for all the advantages of irrigation and protection from floods which the canals confer on him, and that, even after a revision of settlement, Government has not been able to propose a greater enhancement of this rate than 4 annas.

Value of canals.

A fair test of the value of irrigation to a district is to be found in the increase of the rent obtained for the land. On this subject we have fortunately very detailed information in the final report on the survey and settlement of Orissa by Mr. S. L. Maddox, I.C.S. Mr. Maddox states:—"There is little, if any, evidence of general enhancement of rents on the ground of irrigation or of higher rates in irrigated than in unirrigated villages, though there is evidence that rent-rates have risen more in the protected and irrigated tracts than in the unprotected and unirrigated. There is, however, some reason to think that irrigation causes the lowest rents to rise, and in fact has a tendency to equalize rents through an irrigated area. The increase of cultivation is certainly no greater in the protected and irrigated group, and all the enquiries made have failed to elicit any evidence of a substantial extension of cultivation to lands which but for the canal water were not likely to have been reclaimed. Amidst the mass of conflicting information on the subject of the increase of rent-rates, one fact alone can be held to be abundantly proved, and that is that the cases in which a zamīndār has openly enhanced rents on the ground of the accessibility of canal water or has imposed an irrigation cess of his own are very rare." Elsewhere in India the rents of irrigated are sometimes two or three times those of unirrigated land. Here, however, despite increased crops and assurance against drought, the tenant declines to pay a higher rent, and will only pay a very small water-rate for the privilege of irrigation; nor has it been possible to increase this rate at settlement by more than four annas. It must be inferred then that the profit due to irrigation has been very small.

It must not, however, be assumed that these canals are of no value to Cuttack. Apart from any increase which it may have caused in rentals, the canal system is of great value as a security against loss caused by floods and drought. Formerly the landlords' income was precarious because their tenants were liable to failure of crops from these causes. Now there is no such uncertainty, as in the area embanked and provided with distributaries there is protection against devastating floods, and water is available in time of drought for whoever needs it. In the 36 years prior to their construction, 1831-32 to 1866-67, floods sufficiently disastrous to necessitate remissions of Government revenue occurred eight times. The remissions made on this account amounted to over 8 lakhs, and how small a proportion this bears to the total loss suffered in these eight years of disastrous flood may be judged from the fact that in the year 1866, the events of which formed the subject of minute enquiry, the remissions were found to represent only 5 per cent. of the estimated loss. The people, we learn, used to be kept on the alert every year for two or three days and nights waiting for a signal to fly to the highest ground available, and were obliged to see their houses washing down on all sides without having any power to save them. The canal embankments now protect nearly 550,000 acres, and even the greatest floods are powerless to devastate all the country. Immunity from famine is, however, perhaps of even greater value than protection from flood. It appears from the report on the inundations of 1866 that remissions on account of drought, amounting to 14¼ lakhs, or about two years' revenue, were granted in Cuttack in five out of the 36 years ending with 1866-67; and as no remissions were granted unless the loss exceeded one-fourth of the produce, it may be accepted that the losses from drought were both more frequent and greater than is suggested by these figures. No such remissions have been necessary since the construction of the canals; and it is certain that if so terrible a calamity as that of 1865 were again to befall Cuttack, the district would be in a far better position to withstand it. To quote the conclusion arrived at by the Indian Irrigation Commission of 1901-03: "there is no urgent need for further famine protection to the plains of Orissa; and indeed there are few parts of India more secure, or in which the value of irrigation bears so small a proportion to its cost."

CHAPTER VII.

NATURAL CALAMITIES.

LIABILITY TO NATURAL CALAMITIES. THE most difficult problem which the administration in this district has to face is its liability to loss of life and property from natural calamities. The rainfall is in most years ample for its needs, but it is precarious, and its early cessation is fatal to the rice crop on which the people depend. In the deltaic tract which forms the greater part of the district, the difference of level between the high and low-lying lands is so slight that, in the event of any scarcity of rainfall, all parts are equally affected. The low lands are not sufficiently below the level of the uplands to retain moisture for any considerable time after the rains have ceased, and in years of drought the crops grown on them do not compensate for the loss of those which may be burnt up on the arid higher levels. A drought is, therefore, liable to affect Cuttack more seriously than those districts where the difference of level between the up-lands and the low-lying tracts is sufficient to cause the sterility of the former to be compensated by the increased fertility of the latter. Since the droughts, however, of 1836, 1837, 1842 and 1865-66, all of which caused more or less distress, and the last of which brought on the great famine of Orissa, large irrigation works have been constructed which yield an ample supply of water, so that the district may be now said to enjoy comparative immunity from famine, even when there is a protracted cessation of the rains. As a matter of fact, of late years there has been no ground for any great anxiety on the score of drought, although the deficiency of rainfall has in several years seriously affected the outturn of crops. The area now regarded as liable to famine is 1,295 square miles with a population of 572,500, and it is estimated that the maximum number of persons likely to require relief in the event of serious famine is 78,000, of whom 62,000 would be provided for by relief works, while 16,000 would require gratuitous relief.

The next great danger to which the district is exposed is that of inundation. The greater part of the tract of country which it occupies is liable to suffer from the floods of the Mahānadī,

Brāhmanī and Baitaranī, as the channels of those rivers are insufficient to carry off the great volume of water which comes down after heavy rain from the table-lands of Chotā Nāgpur and the Central Provinces. These floods arise from sudden freshets of the rivers before they enter the district and not from excessive rainfall in it; and though a low flood does little harm, as it is prevented from devastating the country by the embankments, the high floods which sweep across the rice-fields do great damage to the standing crop, as they generally occur in July, August and September, when the rice is in the first vigour of its growth or is in flower or nearing maturity. It sometimes, though fortunately rarely, happens that the district is visited with the double calamity of flood and drought in the same year, the former occurring in the early part and the latter towards the close of the season.

Less frequent but scarcely less serious damage has been caused by storm-waves on the sea face; and though the low lands are to some extent embanked against the sea water, violent cyclones breach the embankments and cause great loss of life and property. These cyclones are fortunately rare; they are generally generated during the transition periods antecedent and subsequent to the full establishment of the south-west monsoon, *i.e.*, during the months of April and May, October and November. Their most striking features are the great barometric depression in the centre and the magnitude of the storm area. These two causes produce a large accumulation of water at and near the centre, which progresses with the storm and gives rise to a destructive storm-wave, when the centre reaches the shelving coast. It then sweeps inland, and the damage caused is terrible and widespread.

FAMINES.

Early famines.

Previous to the inception of the great Orissa canal system, droughts and famines were of frequent occurrence. Historical records show that terrible famines occurred in the 14th, 15th and 16th centuries, and during the rule of the Marāthās the district suffered grievously from repeated famines. In the memorable famine of 1770 the land lay untilled; rice was not to be had at two seers per rupee; and while the people were dying by hundreds of thousands, the Marāthā soldiery plundered and devastated the country. Four years later another scarcity is said to have occurred, and in Cuttack town rice could scarcely be purchased at 10 annas for the local seer (105 *tolās*). In 1780 the whole country had sunk into such absolute desolation that there was not a single place except Purī and Cuttack which could furnish even one battalion with provisions. In 1792-93 the miserable peasants again experienced the horrors of famine; scarcity followed in 1803; and when the district passed into the possession of the

British the condition of the country was wretched. A large portion of the land has been thrown into waste; many of the people had fled to the jungle, and the population was insufficient to till the fields. Under British administration an era of prosperity has ensued: with an improvement in their material resources, the people have displayed far more staying power in bad years; cultivation has extended; and though there have been frequent droughts, they have only once culminated in famine.

Famine of 1865-66. The years 1806, 1808, 1809, 1817 and 1828 were years of bad crops and scarcity, and in 1836, 1837 and 1842 Cuttack suffered severely from drought; but the only really great catastrophe of the century was the famine of 1865-66. No such calamity had occurred for nearly a century; it had to be dealt with by a body of officials necessarily ignorant of the signs of its approach, unprepared to expect it, and inexperienced in the administration of relief measures; nor were the native inhabitants more aware of what was coming on them than the British officers. The rainfall of 1865 was scanty and ceased prematurely, so that the outturn of the great crop of winter rice, on which the country mainly depends, was reckoned at less than a third of the average crop. Food-stocks were low, both because the quantity exported in 1865 was unusually large and because the people, unaccustomed to precarious seasons, had not retained sufficient stores at home. When the harvest failed, the gravity of the occasion was not perceived and no special inquiries were instituted; while prices long remained so moderate that they offered no temptation to importers and forced no reduction in consumption on the inhabitants, till suddenly the Province was found to be almost bare of food. It was only in May 1866 that it was discovered that the markets were so empty that the jail prisoners and the Government establishments could not be supplied. But the southern monsoon had now begun and importation by sea or land became nearly impossible. Orissa was at that time almost isolated from the rest of India; the only road leading to Calcutta, across a country intersected by large rivers and liable to inundation, was unmetalled and unbridged; and there was very little communication by sea. By great exertions, the Government succeeded in importing about 10,000 tons of food-grain by the end of November; and this was given away gratuitously, or sold at low rates, or distributed in wages to the starving population. But meanwhile the mortality among those whom this relief did not reach, or reached too late, had been very great; and it was estimated that nearly 1,000,000 persons had died. Though the general famine may be said to have come to an end in November when the new crop began to come into the market, great distress still continued in some parts of the country. The rainfall of the year was so heavy as to cause great floods in the river Mahānadī, and though the harvests in the higher lands were excellent, in all the low lands the crop was drowned. Half the district of Cuttack was thus devastated; in January 1867 forty deaths a day from starvation were reported; and the work of relief had to be taken up again. Altogether about 40,000 tons of rice were imported and lavishly distributed; and about half had been disposed of when the monsoon of 1867, followed by an unusually fine harvest, altogether put an end to the famine in 1868. No complete statistics of the numbers relieved and of the expenditure incurred are available; but the mortality was estimated at one-fifth to one-fourth of the population, and altogether nearly 1½ crores was expended in Orissa during this famine.

The preceding summary of the history of the Orissa famine is condensed from the Report of the Famine Commissioners of 1878, but the catastrophe in Cuttack was so great that a fuller description of the way in which it affected the district seems to be required. The rice crop in the year immediately preceding the total failure of the winter crop of 1865 appears to have been a fairly good one in Cuttack. Even as late as August 1865, prices continued easy, and in that month large purchases were made by a French mercantile house at from 30 to 35 seers per rupee. Though the rainfall of 1865 was below the average, the prospects of the crop seem to have been, on the whole, good up to September; but the last heavy fall took place on the 6th of that month, and after the 18th the rains stopped entirely. Up to October rice continued to be tolerably cheap in Cuttack, the crops were generally promising, and even in the middle of the month people still hoped that a timely fall of rain might save the crop. When, however, the middle of October passed without any sign of rain, the alarm became serious, and by the 20th, the whole country was in a panic. The rice trade was stopped; the country ceased to supply the towns, and at Cuttack the bazars were closed. This refusal to sell, which now manifested itself for the first time, was repeated at intervals throughout the famine; and it was symptomatic of its character, as rather due to scarcity of grain than scarcity of money, that each fresh accession of alarm constantly took the shape of stopping sales at the regular marts altogether rather than of mere sudden enhancements of price. Some dealers really had no grain; others were unwilling to sell on the old terms, and were afraid to raise the terms too suddenly; and the remainder felt themselves unable to meet the demands

History of the famine in Cuttack.

which would have been thrown on them if they had kept their shops open when those of others were closed. Hence the dealers followed one another, and general closing movements took place, which were only got over when the supply had accumulated a little and the alarmed public were glad to accept greatly enhanced rates.

By the 6th November the price of common rice in Cuttack was 8 local seers (105 *tolās*) per rupee. Prices steadily increased week by week, and it became apparent that absolute famine must ensue. The irrigation works, by providing employment for thousands of labourers, who were paid partly in money and partly in rice, rendered the distress later in Cuttack than in the neighbouring districts; but even in Cuttack town the official returns of the 12th February showed the price of rice at from 9 to 7 standard seers per rupee. In April the district began to suffer from actual famine and starvation. The pressure was as yet less in the town than elsewhere; but prices reached 6½ to 5½ seers per rupee, at which rate the people could not long survive, and starving objects began to appear. Several private charities were opened, and there was an old-established public charity; but it was not till the end of the month that the Relief Committee commenced regular operations. The distress was aggravated by the failure of the Irrigation Company's rice, which now came to an end; their funds were at the time scant; they did not import more rice till June; and, meantime, food becoming scarcer and scarcer, the relief afforded by their works was greatly diminished.

It is quite clear that by May there was great starvation and suffering, and considerable mortality in the district, though the mortality in the town was not excessive and there were not the famine scenes witnessed in Balasore. Prices went up to 5 and 4 seers in the latter part of the month, or to about seven times the average price of food; and continuing to rise still higher than this, they did not materially fall during the following three months. From the middle of June to the middle of July, the price in the town of Cuttack (when rice could be bought at all) was from 4½ to 3½ standard seers per rupee, or eight times the average price, and in most places rice was not to be obtained at all. The popular urban confidence in stocks yet remaining in hand only ended in more sudden and complete exhaustion and ruin; and in respect of high prices, Cuttack suffered more than any other district station. Rice was dearer for a short time at Balasore, but the most extreme pressure of prices lasted for a longer period at Cuttack than at either Balasore or Purī. Government relief works were not opened till late in the famine; but the works of the Irrigation Company employed 9,290 persons on an average in each of the six months ending in June. In January when rice was procurable, the numbers were at their highest (14,666); and from that month till June, employment being freely offered, and more and more needed, the decrease in numbers was solely due to the want of rice to feed the labourers. In the rainy months July and August the work was for the most part stopped by the season.

By the end of May, the district was discovered to be in a terrible state of famine. On the 27th May, the Commissioner returning from a tour in the Tributary States found the troops and Government establishments on the point of starvation; and on the 28th he sent a telegram begging Government to import rice for the use of the troops and for the jails, to feed labourers on relief works, and to supply food for the starving. Rice, he announced, was procurable with the utmost difficulty, and then only in insufficient quantities, at 4½ seers per rupee; there were only one day's rations in store for the troops; crime was increasing daily; and all public works and relief works were stopped for want of food. A cargo of 3,000 bags of rice was at once despatched to False Point, but great difficulty and delay was experienced in landing and transporting it; and the greater part of the cargo was not received in Cuttack till early in July. After this, several ships loaded with rice arrived in quick succession, and by November about 10,000 tons of food had been imported. Meanwhile, however, the mortality caused by starvation or by disease, directly or indirectly connected with starvation, want and bad food, was very great. Money was spurned as worthless, and prices were constantly merely nominal. Where rice was to be bought at all, as at Cuttack, it reached the rate of five, four and even three seers to the rupee, but in the interior still higher rates prevailed, even to one seer per rupee. Famishing crowds gathered at the feeding-places, and as one officer wrote, "for miles round you heard their yell for food."

In July some centres for the distribution of cooked food were established in the interior, more were established in August, and in September nearly the full number (43) of feeding-places were in operation. Before this, however, the sufferings of the people had been increased by the inundation of all the low-lying lands. The deltaic rivers, swollen by heavy rain, rose to an almost unprecedented height; the embankments were topped and breached in all directions, and the whole of the low-lying country was flooded by an inundation which lasted for an unusual time. The mortality reached its culminating point in the second week of

August, during the heavy rains which preceded and caused these floods. The people were then in the lowest stage of exhaustion; the houseless poor looked in vain for shelter from the rain that penetrated everywhere; the emaciated crowds collected at the feeding-stations had no sufficient shelter, and the cold and wet killed them in fearful numbers. The known deaths from diarrhœa and dysentery and other similar diseases increased greatly, and the unknown deaths must have been still more numerous, for persons could not reach the *annachhatras* or relief depôts, to which alone they looked for support. In most of the low-lying lands, the *biāli* or early rice crop, which would have been reaped in another week or fortnight, was almost entirely destroyed, and the young cold-weather crops suffered much from protracted immersion. Although new relief centres were opened, yet in several cases it was found quite impossible to supply those already opened with rice, owing to the boats from False Point being unable to make way against the powerful current that then came down; and at several centres operations were altogether suspended. The result of this was a great aggravation of the already existing distress; for those who were congregated at the centres found, when the stock of rice ran out, that they were cut off by the floods from other aid, and many died from sheer starvation. In September some relief was afforded not only by the greater extension and better supply of the feeding-centres and sale depôts, but also by the ripening of the small early crop of rice in tracts which had escaped the flood. At best, however, the distress was still but a degree less than before; rice still sold at six, and even five, seers for the rupee; and it is questionable whether the results of previous suffering, and the effect of unaccustomed food on those who were much reduced, did not cause the mortality to be almost as great as ever. Many, who had lived so long, died when they received the meals to which they had long been strangers.

Up to the end of September, the quantity of rice imported by ship at False Point was as much as could be utilized with the means at the disposal of the local officers and of the Relief Committee. In October, however, the supply of imported rice was almost entirely stopped, owing apparently to misunderstandings between the local officers and the Board of Revenue; only one ship brought a cargo of rice, which however afforded most opportune relief. The sales of rice to the famished people had to be put an end to for want of grain; but gratuitous relief was not checked, as the stocks in hand happily sufficed to maintain (though with great difficulty) the feeding centres; and though in some parts of the country the stoppage of sales was very much felt, in others the market was somewhat eased in the course of October when some new grain became available. In November the new crop began to come into the market in considerable quantities, and then the general famine may be said to have come to an end. The people returned to their avocations, leaving only the emaciated, the orphans and the widows. Considerable distress, however, still existed in the unfortunate tracts which had suffered a second calamity by the floods of August, particularly in the Kendrāpāra sub-division; and in these, relief operations were continued for some time longer.

Owing to the protection afforded by the irrigation works, no famine has occurred since 1866, though there was some scarcity in 1897 in consequence of a flood of great height and of unprecedented duration followed by short rain in September and an almost complete failure of the monsoon in October 1896. All the great rivers rose almost simultaneously to nearly the highest level on record, overflowing their banks or breaching the embankments. They submerged the low lands, which remained waterlogged for more than a month owing to the long duration of the rise in the rivers; and not only was the crop ruined, but much land was thrown out of cultivation by the deposit of sand. On the subsidence of the floods the cultivators replanted as soon as possible, but the next sowings were sacrificed to drought as the old ones had been to flood. The drought was also of long duration, the rains having ceased at the end of September; in some places the winter rice crop, which is the mainstay of the district, failed entirely, and in others the harvest was very poor. In the event, the outturn of rice was estimated at 7 to 8 annas of a normal crop, and allowing for the stocks in hand and the outturn of the *rabi* crops, the deficit to be supplied by imports was put at 10 lakhs of maunds. The affected area was 1,360 square miles, or more than one-third of the district, with a population of 624,840 souls, but distress existed more or less throughout the whole district, and relief had to be given in all the sub-divisions from the middle of March to the end of September 1897. The District Board started relief measures, but it was soon discovered that the circumstances were not such as to justify expenditure by the District Board or by Government, and the grants already made by the Board were thenceforward replaced by contributions from the Indian Charitable Relief Fund and from the Court of Wards Fund in Kanikā and Kujang. Relief was given gratuitously in the shape of grain and money-doles, except in the Kuhunda-Jaipur circle, where arrangements were made to take from the able-bodied recipients of relief such light work as paddy-husking, **Scarcity of 1896-97** rope-making, cotton-spinning and cloth-weaving. In the end, though there was considerable local distress, very little relief was found necessary, and the total expenditure amounted to only Rs. 18,718.

FLOODS. During last century, Cuttack frequently suffered from inundation, and though defensive works in the shape of embankments along the rivers prevented the devastation of the district, a large proportion of the immense volume of water concentrated on the delta periodically spread over the country. Since 1830 floods of a serious character have occurred no less than 22 times, viz., in 1831, 1834, 1848, 1851, 1855, 1856, 1857, 1862, 1866, 1868, 1872, 1874, 1877, 1879, 1880, 1881, 1885, 1892, 1894, 1895, 1896 and 1900. It would be a mistake, however, to suppose that the floods are always destructive. They undoubtedly do harm in many ways, and the greatest of them have caused widespread havoc and destruction; but provided that they are not of long continuance or of great height, and that they come pretty early in the season, these inundations are productive of almost as much good as harm, as they are usually followed by excellent harvests. In many places the receding waters leave a fertilizing deposit of silt, which renews the productive powers of the soil and is of much benefit to the crops; and even the highest floods are of service, as their scouring action results in the clearance of silt on a large scale, and thus increases the capacity of the discharge of the various channels. It is only when their duration or height is extraordinary or when they occur so late as to render re-sowing impossible that very serious and widespread damage is done.

Flood of 1855. With the possible exception of the flood of 1834, the highest flood of which we have any authentic record is that which occurred in July 1855, when the Mahānadī rose to an enormous height and the maximum reading (127·13) on the Lālbāgh gauge was recorded. The embankments were breached in no less than 1,365 places, and besides the terrible losses sustained by the people in the submerged tract, 52 square miles were reported as being permanently left waste for fear of inundation. Fortunately, however, the flood fell as quickly as it rose, and though it submerged nearly the whole district, it did not cause nearly as much damage as a protracted flood would have done.

Flood of 1866. So far as the effect on cultivation is concerned, the duration of a flood is almost of more importance than the maximum rise, and the period of the season at which it may occur is of even more importance. For this reason, the flood of 1866 was more disastrous than that of 1855, as though it was not so high, it continued much longer; and to add to the distress, it came at that critical period of the year when the people were relying on the early rice crop to mitigate the sufferings caused by famine. This inundation broke through the Government embankments in 413 places, and of the 35 embanked rivers, not one was uninjured. Out of 90 *parganas* only six escaped from the flood; 642 square miles were submerged during a period varying from 3 to 60 days, the depth of water being from 3 to 15 feet; and a vast population of nearly 700,000 people are said to have been thrust out of their homes. All the crops were destroyed in the parts affected most seriously; property which had escaped the famine was carried away or destroyed, absolutely nothing was saved, and what the drought had spared was engulfed in the wide vortex of water.

Flood of 1872.

The next great flood was that of 1872, which was very nearly though not quite as high as the flood of 1855. In Cuttack 1,135 square miles were inundated, and in Purī 1,070 square miles; of these 2,205 square miles, about 600 miles were under water for 14 to 17 days, and the rest from 7 to 10 days. Cuttack town was cut off from all communication with the surrounding country and was in imminent danger, being only saved by the energy of the local officers. The canals and embankments were seriously injured, and the country was a vast sea of water stretching from Cuttack to the coast, dotted here and there with a few village sites, where the people and cattle found a temporary shelter and huddled together in the greatest distress. Fortunately, this flood occurred early in the season in the first week of July, and as the subsequent season was favourable, a good crop was obtained from most of the land inundated, and no marked distress was caused.

Flood of 1892.

In more recent years the most serious floods have been those which occurred in 1892 and 1896, the first of which was remarkable for its intensity and the second for its long duration. The flood of 1892 was due to the Mahānadī being swollen to a great height by heavy rainfall, the level at Narāj on the 26th July being 92·10 and at Bellevue on the Kātjurī 88·30. The level above which the Mahānadī may be considered to be in high flood is 88·00 at Narāj, and the river was above this level for only 5 days as against 8 days in 1872. Consequently, the injury to the crops generally was not very serious; and though the embankments in the Purī district were breached in all directions, the damage done in this district was not nearly as great. The Kendrāpāra canal was, however, breached in the 19th and 20th miles, and the Kendrāpāra Extension canal from Mārsāghāi to Jambu was also overtopped and breached in many places.

Flood of 1896.

In the flood of 1896 the Brāhmanī and Baitaranī as well as the Mahānadī rose to a great height almost simultaneously, but the main feature of the flood was its long duration, which exceeded that of all the floods of which we have any record. For 15 days, *i.e.*, from the 24th July to the 7th August, the Mahānadī was continuously above the level of 88·00 at Narāj, except for a few hours on the 2nd August; and on the 25th July it attained its greatest height, 92·10, or the same height as was reached by the floods of 1872 and 1892. The embankments were breached in numerous places, and the Kendrāpāra Extension canal, standing out like the bank of an inland sea, suffered much from the erosion of the waves breaking on it, and was again wrecked. In almost all the parts unprotected by embankments the heavy floods destroyed the *bhadoi* and winter rice crops, and extensive tracts lay under deep water for many days. Some lost their lives and property, and there was considerable distress in all the country open to the ravages of the flood.

CYCLONES.

Cyclone of 1885.

The most terrible cyclone from which the district has ever suffered was the False Point cyclone of 1885, the memory of which is still fresh among the people. It presented two peculiar features, as it occurred during the monsoon months and was of very narrow area, though of unusual severity. The cyclone burst upon the coast in the early morning of the 22nd September 1885, the barometer falling to 27·135″ at False Point Light-house, a reading unprecedented at the level of the sea. It was accompanied by a storm-wave rising to a height of about 22 feet above mean sea-level, which at once submerged the village of Jambu at the terminus of the Kendrāpāra canal to the north-west of False Point, and then rolled on in a north-easterly direction till it lost itself in the Brāhmanī river. The storm was most keenly felt in the Jājpur and Kendrāpāra sub-divisions. In the former sub-division, no less than 2,447 villages were affected and nearly 50,000 houses were destroyed; about 300 human lives were lost by falling trees, walls and homesteads, and 2,973 cattle were killed. The Executive Engineer's house at Akshuāpadā was entirely wrecked, the roof bodily carried away, and some of the masonry pillars destroyed; the Europeans (one a lady), who were in the house at the time, were driven outside, and were for some hours exposed to the violence of wind and rain. In the Kendrāpāra sub-division about 5,000 persons were drowned and 10,000 cattle were lost, 7,000 of these belonging to the Kaldip and Karara *parganas*. These *parganas* included 290 villages with a population of about 26,000 persons, and suffered more severely than any other parts of the district, a total area of about 250 square miles being submerged. Eleven villages were completely swept away, every man, woman, and child being drowned by the storm-wave, and all trace of the houses being washed away; while about 150 more villages were levelled to the ground, though a considerable part of the population managed to escape. The land lying between Rājnagar and the sea face, which before the cyclone was perhaps one of the best rice-growing tracts of the Kanikā estate, was converted into a brackish waste; and in Kaldip all, and in Karara three-fourths of the crops were completely destroyed. By far the greatest havoc, however, was caused on the sea face; here the storm-wave sweeping over False Point Harbour, knocked down all the houses before it, and completely submerged Jambu as it rolled on in an unbroken wave over Kaldip and Karara. The effect of this wave was suddenly to create a sufficient depth of water all over the harbour to float large steamers over shoals where ordinarily there is a depth of only a few feet of water. The sudden fall of the water landed the ships and steamers which had drifted from their moorings on the shoals; while the cargo barges were deposited in the midst of the jungle and in the most extraordinary places, the boatmen having no command whatever over their boats, and being unable to distinguish, amidst the wild waste of water, the creeks from the submerged land. At Jambu itself, out of a population of 130 souls in the village, only about a dozen were saved; the village site, when first visited, was covered with the corpses of men, women and children, while the dead bodies of cattle and deer were floating in great numbers in the creek before the village. Between Jambu and the Brāhmanī, all along the Hansua creek, the scene was one of perfect desolation, with trees uprooted and houses crushed into a confused mass, and with hardly any signs of animal or human life whatever.

Immediately after the disaster the Commissioner and officers of the district staff visited the devastated country and distributed food to the survivors. Relief measures were at once started, depôts being established at Hansua and Rājnagar, at which charitable relief was administered to about 8,000 persons daily. A grant of Rs. 20,000 was made by Government for this purpose, and another grant of the same amount was sanctioned out of the funds of the Kanikā Ward's estate for charitable donations to the ryots. The villages which were not utterly destroyed recovered from the effects of the storm with remarkable rapidity. Trade was for a time suspended, whilst the inhabitants set to work to repair their homesteads; but within a short time few vestiges of the destructive character of the storm remained.

In many of the villages, however, some of the distinctive castes were completely exterminated; so that there are now several *bastīs* in which there are no members of those castes whose presence and services are indispensable to a village community.

Cyclone of 1890.

The last disaster of this kind which has visited the district was in 1890, when a storm-wave affected a considerable part of the Kujang estate, which, since the tidal wave of 1885, had been more or less subject to the inrush of sea-water every year. Matters, however, reached a climax in June 1890, when another wave passed over this portion of the estate, completely destroying the crops in a tract extending from the Keābāg to Gaguā. Here the standing crops were swept away, all the tanks and wells were filled with brackish water, and the supply of food-grain was all but exhausted. Four months after this tidal wave, it was reported by the Manager of the estate that nearly nine-tenths of the people in the affected tracts had no grain in their houses, and had no means of purchasing it. The severity of the distress was all the more keenly felt, as the people had already sustained grave losses in the previous year, when rinderpest carried away more than 75 per cent. of their cattle.

CHAPTER VIII.

RENTS, WAGES AND PRICES.

THE rents paid by the cultivators vary throughout the district according to the quality of the soil they till. The common rate for average rice land is Rs. 3-2 an acre, but good river-side land with a rich deposit of silt, on which tobacco and other valuable crops can be grown, pay as much as Rs. 12 to Rs. 25 an acre; and on the other hand inferior land producing a coarse pulse pays less than a rupee per acre. CASH RENTS.

The rents fixed at the last settlement were not, however, based on the classes of soil under cultivation, as it was found that the villagers were not able to point out, with any degree of accuracy or certainty, definite tracts of lands bearing a uniform rent; and, in these circumstances, the attempt to classify soils by the aid of existing rates of rent was unsuccessful and had to be abandoned. Eventually, it was decided to assess rents on the basis of the existing rates; the basis of the proceedings adopted being the proposal of a fair rent by the Settlement Officer, whether the existing or an enhanced rent for the acceptance of the tenant, the immediate settlement of that rent if accepted by him, and the formal settlement of a fair rent under the provisions of the Bengal Tenancy Act in all cases in which the tenant proved unwilling to accept. The method was simple and involved as little disturbance of the *status quo* as was possible; it avoided the dangers inseparable from systems of soil mapping and the risk of inequalities of assessment due to the idiosyncracies of individual officers; and it was open to any person dissatisfied with the rent proposed to apply for the settlement of a different rent, and to adduce formal evidence in support of his contentions. Under this system, fair rents were settled for the whole body of tenants in the temporarily-settled estates; altogether 650,600 holdings with an area of 822,500 acres were thus dealt with, the average rent throughout the district being Rs. 2-8 per acre. Settlement of rents.

At the settlement of 1837 the average rent of the *thāni* ryots, *i.e.*, the resident cultivators, who held 21·5 per cent. of the assessed area, was Rs. 2-13-3 per acre, while the incidence of the rents of Rents of *thāni* and *pāhi* ryots.

the *pāhi* or non-resident ryots, who in most cases were practically tenants-at-will, was Re. 1-14-6. The area of the latter holdings was over 43 per cent. of the whole assessed area, and the rents paid by them, which have always been more or less competition rents, reflect with much accuracy the general conditions obtaining at the time. With regard to these two classes of tenants, who form the bulk of the cultivators of the district, it was decided that the existing rents of the *pāhi* ryots should as a rule be taken as fair and equitable, and that, if the *thāni* rent was less than the village rent, it should not ordinarily be enhanced by more than half the percentage of the difference between the two. The village rate was calculated on the basis of the *pāhi* rents, and the latter, being competition rents, were thus accepted as the standard of the limit of enhancement.

***Thāni* ryots.** The *thāni* holdings having been held at the same rent since the date of the last settlement, the rents were enhanced wherever they were found to be lower than the *pāhi* rents; but owing to the over-assessment of *thāni* lands in several tracts, it was frequently found that even competition rents had not risen above the rents fixed for them during the currency of the settlement. In these tracts no general enhancement was made, but the excess area was everywhere assessed to rent at the village rate, after making an allowance of 10 per cent. to cover any excess due to the greater strictness of our system of measurement. It is a significant fact as showing at what high rates the rents of *thāni* ryots were fixed at the last settlement, that the general incidence of the rents of *pāhi* ryots did not, even in the course of 60 years, rise to that shewn by the *thāni* rents fixed in 1837. This would not in itself be conclusive evidence as to the over-assessment of *thāni* lands at the previous settlement, as it is to some extent true that these privileged tenancies frequently included the best lands in the villages, but it is borne out by the figures which show how materially the area held by *thāni* ryots and the rents paid by them diminished in spite of the valuable privileges conferred by the *thāni* status. In the event, the average incidence of the rent settled for *thāni* holdings was Rs. 2-14-5 per acre, and for *thāni-pāhi* or mixed holdings Rs. 2-10-10, the enhancement on the existing rent being 6 and 5·4 per cent. respectively.

***Pāhi* ryots.** The general development of the district had naturally been accompanied by an increase in the rents of the *pāhi* ryots, the only class for whom rents were not fixed for the term of the previous settlement. The zamīndārs had consequently enhanced them considerably during the currency of that settlement; and as they were held throughout the new proceedings to be competition rents, they were not liable to any general enhancement. They were, however, enhanced on the ground of excess area or where any particular rents were found to be unreasonably low, either through collusion or fraud, or because they were specially granted as beneficial rents by the zamīndārs. The average rent finally fixed was Rs. 2-13-7, the enhancement on the existing rent being 5 per cent.

Among other classes of tenants whose rents were settled may be mentioned the *jamābandi kharidādārs*, *tankidārs*, *chāndinādārs*, *nisfi-bāziāftidārs* and *kāmil-bāziāftidārs*. The *jamābandi kharidādārs* are the holders of land which, in theory at least, was formerly reclaimed. At the previous settlement they were treated as subordinate proprietors, their rents being calculated at a certain percentage of the assets they received; but at this settlement they were dealt with as tenure-holders under the Bengal Tenancy Act, and their rents were settled accordingly, a percentage equal to that granted to them at the previous settlement (20 to 30 per cent. of the full rent) being deducted from the gross assets of each tenure. The incidence of the settled rate per acre was Re. 1-8, the enhancement being 62 per cent. The rents of the privileged tenants known as *tankidārs*, or holders of small areas permanently assessed at a quit-rent, which averaged annas 3·2 at the previous settlement, were not disturbed, but the excess areas held by them were treated in the same way as those held by revenue and rent-free holders, *i.e.*, they were assessed to rent and settled with them as ryoti holdings. No general enhancement was made of the rent of ryots holding *chāndinā* tenancies, *i.e.*, the holders of homestead lands, the incidence of which was Rs. 3-15-10; but excess areas were assessed to rents at special rates, thus causing a nominal enhancement. The *nisfi-bāziāftidārs*, or holders of resumed rent-free lands, had been assessed at the previous settlement at half rates, and the *kāmil-bāziāftidārs*, or holders of resumed rent-free lands, at full rates. Both these classes were dealt with as ryots whose special privileges had expired at the last settlement, but in consideration of the very low rents at which they were holding, and to prevent the hardship which would have been caused by too sudden an enhancement of those rents, a strict limit was imposed on the enhancement; and they were eventually assessed to rents much below those paid by *thāni* and *pāhi* ryots. The incidence of the settled rent per acre in the case of *nisfi-bāziāftidārs* was Re. 1-1-5 and of *kāmil-bāziāftidārs* Re. 1-5-4, the enhancement being 189 and 55 per cent. respectively. The increase of the rents of the former, which is *primâ facie* extremely large, is due to the fact that these people, who had been given holdings at what were supposed

Other tenants.

to be half rates, were found to be paying what were really peppercorn rents. The rents fixed by the settlement officers are but little more than a rupee an acre, which is less than half the market value of the land.

General result of the settlement of rents.

It was obvious from the circumstances of the case that the rents of a very large number of tenants holding at privileged rates, which had remained untouched for 60 years, would have to be enhanced, while it was known that cultivation had largely extended during the same period, and that there were considerable areas in the possession of both landlords and tenants that had never been assessed to rent or revenue. The general result of the settlement of rents was that the existing rents were retained in 54 per cent. of the holdings dealt with; in 13 per cent. an enhancement was made on the ground of excess area; in 24 per cent. an enhancement was made on other grounds; while 3 per cent. were reduced for loss of area, and 6 per cent. were tenancies newly assessed. The settled assets were altogether Rs. 20,72,900, or Rs. 2-8-3 per acre, as compared with Rs. 18,47,400, the assets existing before the settlement, and the increase was thus 12 per cent. This increase was, however, mainly due to the valuation of land held on nominal payment, the rentals of the *nisfi-bāziāftidārs* being raised from Rs. 35,700 to Rs. 1,03,200. The enhancement actually imposed on the important class of *thāni* ryots amounted to only 6 per cent. The rents of these cultivators had suffered no change for a period of 60 years; and though a large number disappeared in the famine years of 1865-66, it has been held that this is no reason why those who held their ground should be absolved from contributing some share of the large rise in the value of their produce which has resulted from State-constructed improvements and settled government. As already explained, they had from the first been paying highly, and it was considered that their existing rents were fairly adequate; but evidence to the contrary appears to be afforded by the fact that there was an enormous increase in the number of transfers of these holdings as soon as the settlement proceedings were commenced, and that since their close sales and mortgages of holdings and portions of holdings have become very common.

PRODUCE RENTS.

Rents in kind are still paid for a certain proportion of land. The commonest form of produce rent is that known as *dhulibhāg* (literally, a sharing of the dust), which implies an equal division of the grain as well as of all bye-products. Under this system the entire cost of cultivation is borne by the tenants, and when the crop comes to maturity, it is reaped in the presence of the landlord's agent and is carried by the tenant to the threshing floor, where an equal division is made in the presence of both parties. Sometimes, however, instead of the crop being actually divided, it is appraised on the ground, and half the estimated value in cash is taken by the landlord as his share. A less common form of produce rent is that designated *phalbhāg*, *i.e.*, a division of the fruits and grain only, the straw and other bye-products being retained by the cultivator. The statistics obtained during the last settlement for 450 villages show that these systems of rent payment obtained in 4,563 acres, or 1·7 per cent. of the area; and from this it may be deduced that in the whole district rents in kind are paid for about 20,000 acres. It is estimated that in the case of ordinary rice land, the landlord's share is about 8 maunds of paddy, worth Re. 1 to Re. 1-4 per maund at harvest, so that the rent actually paid would be equivalent to Rs. 8 to Rs. 10 in cash.

Besides the *dhulibhāg* and *phalbhāg*, there are two other classes of produce rents, the *panidhān* and *sanjā*. The former is an arrangement by which a portion of a cash rent is payable in kind, *e.g.*, a tenant with a nominal rent of Rs. 4 may have to pay Rs. 3 in cash and Re. 1 in grain. The landlord fixes the rate, so that the tenant generally has to pay something more than he would obtain for his grain in the open market. *Sanjā*, *i.e.*, a contract, is a term applied to the payment of a fixed quantity of agricultural produce. The latter generally amounts to about 6 maunds per acre, which would fetch Rs. 6 in a good year and Rs. 9 in a bad year. The quantity fixed has to be paid whether the season is favourable for the ryot or the reverse, and the rent thus presses most heavily on him when he is least able to afford it. Fortunately this system is very rare in Cuttack.

Statistics of the wages given for certain selected classes of labour and the rates current for the decade 1893—1902 will be found in the Appendix. It is of some interest to compare these wages with those current in 1805, when sawyers, stone-cutters and tailors got Rs. 4, bricklayers, blacksmiths, carpenters, thatchers and syces Rs. 3, and grass-cutters, sweepers and other menial servants Rs. 2 a month. If hired by the month, ordinary male labourers received Rs. 2-8 and women and boys Re. 1-14; if paid by the day, the rate was 1 anna 10 pies for male coolies, and about one anna or a little less for boys and females. Sir W. W. Hunter found that between 1850 and 1875 all wages that were paid in money had risen by more than one-third, and that all wages paid in kind had remained the same. "The wages of agricultural day-labourers," he wrote, "are generally paid in kind, and do not seem to have altered since 1850. Such wages were then about twelve to fifteen pounds of unhusked rice per diem; WAGES.

and the same rate continues at the present day, except when they are calculated according to the money value of the labourer's hire. All labour, however, paid by money wages has increased in price; and in the large towns, such as Cuttack and Jājpur, field work is now frequently paid in this way. Agricultural labour is always paid at a lower rate than other unskilled work. Day-labourers, other than agricultural, now receive from 1 anna 4 pies to 1 anna 6 pies in the rural tracts, and 2½ annas a day in the towns; in 1850 the wages were three-fourths of an anna in the rural tracts, and 1 anna 6 pies in the towns. Smiths and carpenters now get 2½ annas in the country and 4 annas in the towns; in 1850 their wages were 1½ annas and 3 annas respectively. Bricklayers, who are only employed in the towns, earned 1½ annas a day in 1850, and now receive from 3 to 4 annas. On the whole, it may be said that labour fetches double in the towns what it does in the country; and that, during the last twenty-five years, from 1850 to 1875, the rates of wages have risen from thirty-five to forty per cent."

On the whole, there has been a rise in the price of labour during the last 30 years, owing largely to such causes as the extension of the Orissa Canals, and more recently to the construction of the railway. The upward tendency is more noticeable in the towns than in the villages, and the increase of wages is more marked in the case of skilled than of unskilled labour. Away from his village, a carpenter now gets 6 to 12 annas a day, and a good blacksmith will not work for less than 8 annas and can even command 12 annas a day; while ordinary male day-labourers earn 2½ to 3½ annas, and women and boys 1½ to 2 annas per diem. In his own native village, a skilled labourer gets from 4 to 6 annas, and an adult unskilled labourer from 1½ to 2 annas a day; but the amount of the wages paid depends on the demand for labour, the nature and amount of the work done, and the size and position of the village, *i.e.*, whether it is in a remote and out-of-the-way tract or in the neighbourhood of a town. For making and repairing agricultural implements, carpenters and blacksmiths, who are still an essential part of the village community, are always paid in kind, the annual payment averaging about 9 seers of rice from every client; and when paid in kind the day-labourer gets varying quantities of paddy equivalent to 2 to 2½ seers of rice. Measured by the quantity of grain given, there does not appear to have been any increase in the wages paid to agricultural labourers during the last 30 years; but owing to the enhanced price of food-grains, the money valuation of wages in kind has increased by 90 per cent. On the other hand, though the wages paid in cash have increased considerably, they have not risen in the same proportion as the prices of the staple food-crops; and the condition of the town labourer has deteriorated, while that of the village labourer who receives his wages in kind has distinctly improved. Money wages are now slightly less in value than wages in kind; the latter are, therefore, always preferred by the village labourers, and it is extremely difficult to obtain a cooly in the mofussil who will work for cash wages in the sowing and reaping seasons, when wages in kind are freely given.

A statement of the prices current in each sub-division during the years 1893—1902 will be found in the Appendix. The enormous rise in the price of the staple food of the country which has taken place in the last century is sufficiently demonstrated by the fact that between 1811—1817 the price of rice averaged from 30 to 41 local seers (105 *tolās*) to the rupee according to its quality, while the average price of paddy was 90 seers to the rupee, and that even as late as the 10 years ending in 1856 the average price of common rice was as low as 58 standard seers (80 *tolās*) per rupee. There was then a sudden rise; in the next decade it was only about half as dear, being sold for 28½ seers to the rupee; and in the succeeding decades the price rose steadily to 27, 21 and 18 seers. The list of prices given in Toynbee's History of Orissa (Appendix III) shows that in the last 100 years there has been a very great rise in the price of agricultural produce, pulse, *ghī* and tobacco; cotton yarn and oil have cheapened, but on the other hand there has been but little change in sugar, salt, and the betelnut which every Oriyā chews. Thus the cultivating classes gain both on the better price they obtain for their surplus produce and the smaller price they pay for imported luxuries. The labourers in the villages have been equally benefited by the rise in the prices of food-grains, as wages are still paid in a great majority of cases either wholly or partly in kind. This system is particularly suited to an agricultural district like Cuttack, as it has the advantage of being unaffected by any rise in the price of food-grains; and the result is that whatever fluctuations may take place in the market, the labourers' wage remains the same. PRICES.

The following account of the material condition of the people is taken from Mr. Maddox's Report on the Survey and Settlement of Orissa, 1890—1900:—"It is very difficult indeed to get any accurate understanding of the material prosperity of the people at large. Enquiries at once put the villager on his guard, and he makes such answers as he thinks will conduce to his gain, without any great regard for the truth. The general opinion of the officers MATERIAL CONDITION.

who have for years worked among the people and gained their confidence is that 80 per cent. of the rural population are more or less permanently indebted to the *mahājan*, proprietary tenure-holder, or zamīndār. The remainder are themselves landed proprietors, or have other means of maintenance than agriculture. It does not, however, follow that because the villager is in debt and has no capital or savings to meet the strain of a bad season that he is very much to be pitied. The *mahājan* is almost always a local man, and generally the tenant cultivates the fields over which the *mahājan* has a lien. Again, the proprietary tenure-holders are local men, and frequently have stocks of grain. Advances of grain for food and seed are not infrequently given by zamīndārs, especially in times of scarcity, and are repaid by return of the principal with 25 per cent. interest when there is a sufficient crop. Such conditions, however, become oppressive when for two years in succession there is scarcity. The Oriyā cultivator is content with very little, and that he generally gets. A full meal of rice once a day, taken with a little salt, some pulse or vegetables, and perhaps fish, suffices him, and he eats cold in the morning what is left over from his evening repast. Animal food is a luxury, but well-to-do men eat a little mutton and goat's flesh, and all classes eat game whenever they have the luck to kill any. The poorest classes take, to supplement their rice, boiled *kulthi* and *mandiā* cakes, and find a substitute for vegetables in the many herbs and grasses that grow wild, and it is very few indeed who cannot fill their bellies with food which, if not appetising, is certainly satisfying. If the harvest fails or supplies run short, the cultivator finds in the *mahājan* a banker always ready to advance money on good security, and able and willing to tide him over hard times, provided there is no abnormal general distress; and the history of the floods and drought of 1896 shows that the agricultural community can withstand very serious calamities, if the bad season is followed by a good harvest in the next year."

The difference between the two years 1865-66 and 1896-97 is very striking, and clearly illustrates the great improvement in the resources of the people which has been the result of the changes effected during the last half century. The years preceding the famine of 1866 had been a most prosperous era, distinguished by a great development of the exports, and the terrible distress that followed the drought of 1865 seemed to disprove the truth of the popular belief that the food-stocks left after a good year are sufficient to meet the wants caused by a single harvest's failure. The years 1890—95, on the other hand, were marked by failure of rain and floods in 1891-92, and by heavy exports in subsequent years, which might have been expected to deplete the stocks in reserve; yet not only did the district escape famine, but it exported over 16¼ lakhs of maunds of grain in 1896-97, though there was a great flood which destroyed the crops early in the year, followed by short rain in September and an almost complete failure in October, which resulted in a very short crop of the *sārad* rice on which the people mainly depend.

Dwellings. The actual requirements of the people are very few. Brass and bell-metal utensils and ornaments, coarse cotton cloth, and certain other articles manufactured for local consumption supply nearly all their wants, and they need very few such articles of foreign manufacture. The one mild luxury which is practically universal is that of smoking; the tobacco is smoked not in the hookah, but in the form of cigars, or still more economically, by being pulverized and wrapped up in a leaf. The house of an ordinary ryot is one of which the walls are made of mud, dug from some spot close by, with a roof generally consisting of a framework of split bamboos tied together by string and thatched over with straw or grass. The floor is smoothed over almost every week with cow-dung and mud, and the walls are decorated with fantastic figures of crude design and glaring colours. Even the house of a substantial ryot, comprising a sitting room, a bed-room for males and another for females, a treasure-chamber, a cowshed and a sanctuary for the family idol, besides a large court-yard, verandahs, etc., would not cost more than Rs. 100. The furniture is equally simple, and the articles which the cultivator requires for himself are extremely few. Most of the furniture in the house of an ordinary ryot are articles required for his work, such as mats for drying paddy, and baskets and earthen jars for holding grain. For his own comfort he has merely a few mats for sleeping on, a basket in which to keep his clothes, and a small plank to serve as a seat; though mats give place to wooden beds and cane baskets to boxes of wood or tin in the houses of the wealthier ryots. On the other hand, the peasant is anxious to have a good stock of kitchen utensils and buys as many as he can afford; but all the vessels necessary for domestic use, whether of brass, bell-metal, iron, wood or earthenware, can be purchased for Rs. 12, and even a well-to-do ryot can get all he wants for Rs. 50.

Dress. He is equally eager to give his women folk as many ornaments as his purse can supply, and the female members of his family deck themselves out with a large variety of bracelets, armlets, anklets, hairpins, and rings for the ears, fingers and toes.

But though the number of ornaments worn by peasant women is very large, their intrinsic value scarcely exceeds Rs. 5, as they are generally made of cheap materials, such as glass, lac, bell-metal and brass. It is only the wealthier ryots who can manage to give their wives gold and silver ornaments, and the value of the jewelry they wear is about Rs. 80.

Though keenly alive to the advantages of jewelry as a setting for their charms, the women are content with a poor wardrobe. The universal garment is the *sārī*, a long piece of cloth draped over the head and shoulders, which is tied at the waist and reaches down to the knees. They generally have only two of these cloths, costing about Rs. 3; and though a substantial cultivator's wife has more cloths of a better quality, the total expenditure is not more than Rs. 30. The man himself generally wears a *dhotī* and a *chādar*, or loose cotton sheet, worn over the shoulders; and in the fields he is content with an exiguous rag, called a *gāmchhā*, round his loins, and with a wicker shield-like hat to protect him from the weather. His outlay on clothes is not extravagant, as an average peasant's outfit both for ordinary wear and for festive occasions costs less than Rs. 4, and his rich neighbour's clothes can be got for Rs. 30.

Indebtedness.

On the subject of the indebtedness of the ryots, Mr. Maddox writes:—"If we except well-to-do ryots holding themselves 8 or 10 acres or more of land, there are few cultivators not in debt to the *mahājan*. Enquiries made by the Assistant Settlement Officers have not resulted in much definite information, but it appears to be well established that the petty cultivator pays away as rent and interest the whole of his crop, except what is left to him for subsistence allowance, and has almost always to borrow again before the next harvest is ripe. In a good year he will pay his debts and rent in full in January or February, and have enough to carry him on throughout the year, if he has no unusual charges to meet, but he will probably seize the opportunity to marry off a son or daughter, which may leave him in debt to the extent of Rs. 50 for which he will probably execute a mortgage on his land or a portion of it. In bad years payment of rent and loans will not leave the cultivator enough to eat, and he will probably borrow again in June or July enough to carry him on to the end of the year. Most of the money, or rather grain, lending in Orissa is in the hands of zamīndārs or proprietary tenure-holders, and they are by no means hard creditors. Very occasionally a bad tenant gets sold up, but as a rule the zamīndār knows that it is to his interest to keep his ryot going, and will supply him on the usual terms with the necessary grain, even though there be a large accumulated debt. As security he nearly always requires a bond pledging the land, but these are not often used, and sometimes even particulars of the land and of the consideration are not filled in. The only matter in which I have known ryots to show any providence is that of preserving grain for seed, and several instances were brought to my notice in 1897 where the cultivator, though reduced to great straits for food, getting practically no rice to eat, refused to touch his little seed-store."

Condition of the agricultural classes.

In spite, however, of their indebtedness and of the liability of their crops to injury from droughts and floods, the agricultural classes have more resources than any other. Not only have they better means and better credit than the labouring classes, but being in the habit of keeping grain for home consumption, those who have crops of some kind are in a better position than the non-agricultural classes when grain is scarce and prices are high. In the course of the last settlement it was ascertained that the average area of a holding in Cuttack is $1\frac{1}{4}$ acre, which is at first sight a very small amount; but a ryot often holds land in more than one village, and from the statistics of 25 typical villages, it was found that out of 100 heads of families, 56 held over 2 acres each, 30 held 2 acres or less, and 14 had only homesteads. The latter were for the most part labourers; those holding 2 acres or less were artisans, weavers and others, who combine agriculture with their hereditary occupations; and the average holding of the agriculturist was found to be 3·23 acres, and of the whole population 2·21 acres per head of a family. It has been calculated that a quarter of an acre of homestead land with 3 acres of irrigated or $3\frac{1}{2}$ acres of unirrigated land would support the ordinary agricultural family of five on a subsistence allowance of food; but the regular income derived from the land is supplemented by the income obtained from subsidiary occupations as well as from the remittances sent home by those members of the family who live out of the district, most of whom are in service and earn good pay. The statistics which have been compiled show that in any ordinary year there must be a considerable excess of production over consumption, and that, after allowing for the amount required for seed, the outturn of food-grains leaves nearly three-fourths of a seer per head.

Condition of the labouring classes.

As regards the labouring classes, the village artisans who never go out of the village form a recognized part of the village organization, and are indirectly supported by agriculture. The ordinary artisan with a family of five earning 7 annas a day does not spend more than 5 annas, and is thus able to lay by something which enables him in time to invest his savings in land, the great ambition of every man in Orissa. There is hardly any really skilled artisan, who has not, if he is a man of the mofussil, some land, and if a man of the town, some money-lending. Here, however, as in other parts of India, the lot of the day-labourer is rather hard. Spending what he earns from day to day, he has very little to pawn or sell in times of distress, and he is therefore the first to succumb in time of scarcity unless he is carefully watched and given work within easy reach. Unlike the Bihāri, the Oriyā does not move with his family in search of work, and no labourer cares to go to a distant place for employment leaving his family uncared for and with the prospect of only earning enough for himself. The *kuthiā* or *haliā*, *i.e.*, the unskilled labourer who is engaged by the year and paid daily in kind, is a little better off than the day-labourer. He has a better man to look after him, who, if an old master, does not forsake him till he is himself reduced to the very last straits; and besides this he generally is allowed to hold about half an acre of land as *jāgīr*.

CHAPTER IX.

OCCUPATIONS, MANUFACTURES AND TRADE.

IN Cuttack, as in other parts of Bengal, a large majority of the population are engaged in agricultural pursuits, and the number of those who obtain their livelihood from other sources is comparatively small. Agriculture supports 58·5 per cent. of the population, industries 18·3 per cent. and the professions 2·6 per cent. Of the agricultural population, 31 per cent. are actual workers, of whom 19,000 are rent-receivers, 327,000 rent-payers and 22,000 field-labourers. Of the industrial population, 55 per cent. are actual workers, and these include 15,000 cow-keepers, 16,000 fishermen and fish-dealers, 19,000 rice-pounders, 14,000 firewood- and charcoal-sellers, 33,000 cotton-weavers and spinners, 5,000 gold and silversmiths, 3,000 ironsmiths, 3,000 carpenters and 12,000 basket and mat-makers. Of the professional classes, 37 per cent. are actual workers, including 3,000 priests, 6,000 persons employed in temple service, 1,000 astrologers, 4,000 teachers and 2,000 musicians. Among those engaged in other occupations are 22,000 herdsmen, 13,000 beggars and 98,000 general labourers. Large, however, as is the ratio of agriculturists, this figure does not convey a strictly correct idea of the number of persons engaged in agricultural pursuits. In Cuttack there is scarcely a single caste that does not live by cultivation, even though it may not be the principal or only means of support. Brāhmans, who are interdicted from touching the plough themselves, employ labourers to till their lands; boatmen, fishermen, washermen, barbers, shepherds, cowherds, potters, weavers, oilmen, etc., while carrying on their respective caste avocations, cultivate some land at the same time; and it is probable that the actual number of those who obtain either the whole or part of their income from the soil is considerably larger than the figure quoted above. OCCUPATIONS.

With the exception of the silver filigree work of Cuttack, the industries of the district are of little importance. As is only natural in a district where the great majority of the people are engaged in agricultural and pastoral pursuits and where the urban population is small, the bulk of the industrial community are MANUFACTURES.

engaged in supplying the simple needs of rural people. The people require very few articles of foreign manufacture; and brass and bell-metal utensils, coarse cotton cloth, and certain other articles manufactured for local consumption, such as woollen blankets, paper, pottery, etc., meet most of their wants. In the interior the artisans who make these articles form a component part of the village organization, some of them still holding service lands for their work; and even in the towns the manufactures are mostly of a primitive kind producing little for export.

Gold and silver work.

Cuttack is one of the few places in the Province in which gold and silver work is carried on. This industry gives employment to about 400 families in Cuttack town and to 2,000 families in the interior. The instruments used are of the simplest kind and the equipment of the workshop is equally primitive, as a small hole in the mud floor which serves for a furnace, an earthenware bowl, and a couple of fans without handles generally form the whole apparatus of the workmen. The artificers, patiently working by a trying light in ordinary earth-paved huts with these crude implements, turn out articles of extreme delicacy and finish, and often of exquisite design. The majority are ornaments intended as tributes to feminine vanity, such as *alakās* for the forehead, hair-pins with four or five chains suspended from them, strings of flower-buds worn round the hair, and numerous kinds of ear-rings, some for the lobe, others for the side, and a third class for the upper portion of the ear, besides a number of ornaments for the nose, the neck, the wrists and arms, the waist and ankles. To this list must be added a long catalogue of silver fancy articles and ornaments of filigree work which are usually kept for sale in the Cuttack market. Some of these are vessels used at religious festivals and on ceremonial occasions, such as the rose-water vases or sprinklers brought into use at weddings, nautch parties and other festivals, and the *attardāns* and silver platters for betel-leaf which are handed round at *darbārs*; in addition to these, there are various articles of a more modern character, such as cigar-cases, card-cases, picture-frames, buttons and studs, figures of animals, flower-vases, etc.

The crude material consists of gold with an alloy of silver and copper, and of silver alloyed with copper. The molten alloy is poured into a mould, and the bar thus produced is beaten on the anvil into a thin rod. The latter is then passed through the holes made in the *jantā*, *i.e.*, an iron plate perforated with holes varying in size. The thick wire obtaining by drawing the metal through the larger holes is gradually made thinner by being drawn through the smaller holes, some of which are about the size of a pin's head, while others are no bigger than a pin's point. This process is continued till the wire becomes finer and finer, and the requisite diameter is at length obtained; in this way 120 feet of wire can, it is said, be drawn from a rupee's weight of silver. The wire is then cut up into the lengths required, bent and coiled into various shapes, and then arranged according to the design of the work on a piece of mica placed over a tin plate. When the different kinds of wire have been arranged and the design has been filled in, they are soldered together and the mica and tin plate are removed. The article is then coloured and burnished, a paste of salt and alum being put over it, and after being dried it is ready for sale. The finest articles are produced in the Orissa Art Wares factory, the proprietor of which, Mr. M. S. Das, C.I.E., had brought the improved methods of modern mechanism to the assistance of the hereditary training and skill of the native artist. In purity of design and delicacy of finish the articles turned out in this factory surpass the gold and silver work in every other part of the Province, and the graceful skill with which the spider-web of wire is manipulated has made Cuttack justly famous for this kind of work. No accurate information as to the quantity of gold and silver used is available, but about Rs. 1,00,000 worth of gold and about Rs. 3,25,000 worth of silver are imported into Cuttack annually. The gold and silver ornaments and fancy articles produced are exported not only to Purī and the Tributary States of Orissa, but also to Calcutta, Delhi, Bombay, Cutch, Bangalore, Akyab, Mauritius, and other places.

Salt manufacture.

Formerly the manufacture of salt was a valuable industry, and Stirling has left it on record that the finest salt of all India was manufactured in the wild unhospitable tract along the sea coast of Orissa, and that under the monopoly system the East India Company obtained from this source a net revenue falling little short of 18 lakhs of rupees. In 1897-98 altogether 162,551 maunds of salt were manufactured, but the industry has now been ruined by the railway, which has encouraged the importation of the cheaper Madras salt.

Cotton weaving.

Weaving is another industry which has suffered from competition with imported goods. When the English first came to Orissa, the fine muslins of Cuttack were eagerly bought up by the English factors; but owing to the preference of the people for cheap machine-made goods, the industry is now practically extinct. Weaving of cotton cloths is still carried on at Gulnagar, Jankoti and Kishornagar, and indeed in most of the villages in the interior; but the hand-made article has been driven out of the market in nearly every part f Cuttack by imported piece-goods, and the country looms are almost at a standstill. At the last census only 33,000 cotton-weavers and spinners were enumerated; many of the weaver caste have entirely forsaken their hereditary occupation and live solely by cultivation; and those who still work at the loom are forced to add to their income by taking service as labourers if they have not sufficient capital to rent land for themselves. Their manual labour cannot compete with European machine-made piece-goods either in fineness or cheapness of price. The cloth they turn out, though stronger, is much coarser, and is used only by the poorer classes, who manage to make one piece last throughout the year. Fine cloth is however still woven at Gulnagar from imported twist; it meets the local demand, and also finds its way into the shops of Cuttack, where it is bought by the richer citizens and still commands a fair price.

Silk weaving.

The weaving of *tasar* silk on a small scale is carried on in the village of Gopālpur in the Dharmshāla thāna. The silk industry, like the cultivation of the betel leaf, was introduced by a few Bengalis who emigrated to Cuttack from Burdwān or Midnapore. There are about 200 families engaged in the industry, who are the descendants of the weavers who settled here about six generations back. Almost every family owns a loom, and manages to obtain a fair income without having to take to any subsidiary profession. The process they follow is as simple as the scale of their business is small. They maintain no large filatures, possess no intricate or improved form of machinery, and use only the *tasar* cocoons which they obtain from the jungle tracts to the west. The *tasar* cloth which these weavers produce is of a coarse and inferior quality. The bulk of the outturn is taken by traders in Madras, Berhampore, and other places in Southern India, who make advances which are adjusted when the fabrics have been finished. The value of the cloth purchased by these merchants is reported to be about Rs. 12,000 annually, and a small quantity, valued at Rs. 500 a year, also finds its way to Cuttack.

Paper manufacture.

Paper was formerly manufactured by Muhammadans from the *chanpat* fibre, but the industry has declined owing to the competition with imported cheap paper of better quality. For a long time it lingered on, as the local *kachahris* and offices used the local paper, till it was superseded by the Bally paper. The seats of the industry were Kendrāpāra, Asureswar, Hariharpur, and Padmapur, the last place being the chief source of supply for the Division. The industry would indeed have disappeared long ago were it not that owing to its durability the paper is still in demand in the Cuttack Collectorate, where it is used for the flyleaves of records. The demand is however small, and even the few families which keep up the industry depend very largely on labour or eke out their earnings by trading in sheep and goats.

Other industries.

The other industries are few in number and of little significance. A small colony of stone-carvers settled at Naltigiri about 34 miles from Cuttack make images of gods and ornaments for temples in conventional designs; the industry is said to have been introduced by the Marāthās, but it has made no progress. Locks, nails, nut-crackers, etc., are made at Baroda and Kalāpadar in the Bānki thāna. The carpenters of Cuttack town have some skill in cabinet-making and a certain amount of wood-carving is still carried on. Formerly the temples, *maths* and large houses had the woodwork in their inner verandahs elaborately carved, while the doors and windows contained good specimens of lotus and geometrical screen patterns; but now there is very little carving done, beyond the grotesque designs of gods and fabulous beings which are carved on the panels of doors and on boats. In Cuttack the manufacture of toys and sticks from buffalo horns is a speciality for which there appears to be a growing demand. The horn is cut to the required length, turned on a primitive lathe, and then polished; these articles obtain a ready sale, but the industry is only on a very small scale and is confined to a few families. The only other indigenous industry which appears to prosper is the manufacture of brass and bell-metal utensils and ornaments. This has not yet suffered from the competition of foreign or machine-made articles, though gilt ornaments of German silver are said to find favour with many of the women on account of their lightness, cheapness and gloss. The vessels manufactured are exported to Purī, Sambalpur, Sonpur, Ganjām and Calcutta, and a thriving trade is carried on.

Mines.

There are no mines in the proper sense of the word. Sandstone, laterite and rubble are quarried from the hills along the western border, but only for the railway and local use. The laterite found in the gneissose hills along the road from Calcutta to Cuttack is admirably adapted for building by its durability, and is commonly used for this purpose.

Factories.

The Government workshop at Jobra is the only factory in the district or indeed in the whole of Orissa. This is a civil engineering workshop with steam as the motive power, at which the iron and wood-work required for the anicuts, locks and canals are turned out; it employs an average of 173 operatives daily.

TRADE.

Enligsh factories.

From the fact that the first English factory in Bengal was established in this district, it is clear that it was at one time a place of some commercial importance. The first English expedition to Bengal set forth from Masulipatam in 1633; the party consisted of eight Englishmen, and the good ship, "Hopewell," that bore them was no better than a native junk. With favourable winds the expedition* managed to reach Harispurgarh, but soon after they arrived, a Portuguese frigate came in and anchored near them; and when the English went ashore, the master of the frigate "with the assistance of some of the ribble-rabble rascals of the town" set upon them, and the English "had like to have been all slain and spoiled, but that Lucklip, the *rogger* or vice-king there, rescued them with 200 men." Cartwright, the leader of the little band, soon afterwards set out with a valuable consignment of gold, silver, cloth and spices for Cuttack, where the Muhammadan Governor of Orissa held his court. Here he demanded redress for the Portuguese attack upon them and applied for a permit authorizing the English to trade in Orissa. The Portuguese captain also appeared however, and charged the English crew with fighting in order to make a prize of his vessel and take his goods by force. A nobleman of the court was bribed by the Portuguese and espoused their cause; and Cartwright, finding he could get no satisfaction, "rose up in great anger and departed, saying that if he could not have right here, he would have it in another place; and so went his way, not taking his leave of the *Nabob* or of any other; at which abrupt departure they all admired." At his next interview he found the Nawāb in a far more favourable mood, and told him boldly with a stern undaunted countenance that "he had done his masters of the Honourable Company wrong, and by his might and power had taken their rights from them, which would not be so endured or put up." The Nawāb then made some enquiries about the power and trade of the English and was informed that their maritime strength was invincible. This reply made a deep impression on the Nawāb and finally he gave the English permission to trade in Orissa. On the 19th May 1633 the English left Cuttack for Hariharpur, a village adjoining the present Jagatsinghpur; here they at once proceeded to build a factory, this site being chosen apparently because, to quote Bruton's account, "the town is very full of people; there are many merchants in it and great plenty of all things: here is also cloth of all sorts, great store, for there do belong to this town at least 3,000 weavers that are housekeepers, besides all other that do work, being bound or hired." This factory however was not long maintained, as the merchants did not understand the necessity for severe self-restraint and temperance in a hot climate; the country abounded with fruit and

* For a fuller account of the fortunes of this expedition, *see* the Early Annals of the English in Bengal, vol. I, by C. R. Wilson.

arrack, and these when taken with excess produced lamentable consequences. The place consequently acquired a bad name among the English, and its unhealthiness was one of the most serious obstacles to their progress. Trade was also crippled by the attacks of the Aracanese pirates and the opposition of the Portuguese; and to add to the difficulties of the English, the river where their vessels used to lie having gradually silted up, it became unsafe for ships to ride there and difficult to send goods by sea that way. In 1641 the factory at Hariharpur was found to be on the point of dissolution, only a few fine muslins were in preparation, and the trade henceforward was diverted to Balasore which possessed far greater commercial advantages.

Trade under the Marāthās.

We have little information concerning the commerce of the district during the 150 years which elapsed before the British conquest; but enough is known to show that the trade of the country was paralyzed by the oppressions of the local officials. However cheap might be the inland markets, the tolls and custom-houses on the roads and rivers made the goods too dear for exportation, and anything like internal trade was rendered impossible by the incessant black-mail which was levied. Besides the royal officers who imposed a tax at every few miles, each petty proprietor through whose estate the route lay lined the road with hungry myrmidons; and in the short journey of 103 miles from Cuttack to Balasore the tolls amounted to nearly a third of the total value of the goods.

Trade under the English rule.

The incursions and oppressions of the Marāthās soon put an end to whatever commercial prosperity Orissa may once have possessed, and the trade in rice and salt, which had survived Marāthā misrule, was considerably diminished when the Government asserted its right to the monopoly of the latter article. When we took the Province, however, considerable quantities of these articles were still exported from the ports, which were chiefly frequented by three kinds of craft, viz., (1) Maldive vessels, which brought cowries, cocoanuts, coral, and dried fish, and took back rice and earthen pots; (2) sloops which carried the Government salt to Calcutta; and (3) sloops, built at Contai and Hijilī, which only came in the cold season and carried rice to Calcutta. The Rājā of Kanikā carried on a considerable trade in rice on his own account, and large numbers of swine and horned cattle found their way by land to the Calcutta market. The export trade, however, gradually dwindled down to a fraction of its former importance, and in 1813, the Collector reported that the only articles exported were rice and a little salt (about three lakhs' worth), and that trade was hardly known even by name. The internal trade was equally limited and was confined to the supply of rice and other articles of every-day use or consumption to the larger towns, and to the mutual exchange of surplus produce and articles of home manufacture at the *hāts* or markets in the interior.

Then, as now, the grain markets in the towns were in the hands of middlemen, who so regulated the supplies coming into the town as to keep up prices. When the rivers rose to an unusual height, or the crops failed to any considerable extent, the difficulty which the inhabitants of Cuttack experienced in getting food was extreme. The military authorities were constantly representing to the civil officers the difficulties they had in getting supplies for the troops. They asserted that they had often to go without sufficient food for several days together, and declared that unless this state of affairs was remedied, the sepoys would lose all sense of discipline, and setting their authority at defiance, plunder the grain shops in the town of Cuttack. In consequence of these complaints, the Collector was authorized in 1805 to advance the sum of Rs. 10,000 for the purpose of purchasing and storing a supply of rice for the town of Cuttack and for the military or cantonment bazar. A godown was built for the storage of this large stock, and a public market established for its sale on the site of the present Chāndni Chauk. For many years after this, Cuttack continued to be in constant want of supplies and frequently on the verge of famine. Time after time urgent calls were made on Balasore for rice, and pilgrims had to be warned of the scarcity in the district and directed to supply themselves with provisions before entering it. There was however no scarcity 20 miles north of the Mahānadī; at Balasore rice sold at 65 seers for the rupee and there was enough in store for three years' consumption; there were immense stocks at Dhāmra and Churāman intended for export to Madras; and consequently it was thought that the scarcity of rice was not natural, but must have been artificially produced. It was pointed out that the large number of Marāthās still resident at Cuttack were bitterly hostile to the English, and did their best to stop the import of rice in the hope of starving them out; they themselves ceased to import from Sambalpur, and they prevented the ryots, many of whom were still in their debt, from bringing in grain to Cuttack. This was all the easier, as the ryots had always been accustomed to give up nothing until they were compelled. The Marāthās took what they wanted by force, and the ryots did not understand our method of asking for and paying for what we wanted; they took it for weakness, and were so elated at their release from oppression that they thought themselves quite independent and would do nothing to oblige any one. Besides this, even the *âmils* were in league against the English, as they had for a long time taken advantage of their position to secure the lion's share of the profitable export trade to Madras, and did not wish to sell in Cuttack.

In order to remedy this lamentable state of affairs, land was offered rent-free in perpetuity to any merchants who would establish grain shops near the site of the public market. It was at this time, and in consequence of this offer, that a row of houses arose on either side of Chāndni Chauk, some of which exist in a dilapidated condition to the present day. But though this measure gave some relief to the inhabitants, its effect was only temporary; the usual tightness of the market was enhanced by severe droughts and by floods which cut off communication with the interior; and the officers of Government were constantly interfering with the market in order to try and force down prices. The Governor-General was at last forced to issue stringent orders prohibiting this practice, and in 1821 a proclamation was issued declaring the freedom of trade and the absence of all restrictions affecting the grain market. Since that time trade has steadily developed with the opening of new communications, which have placed nearly every part within easy reach of the markets. The roads have been improved and extended; a great impetus has been given to the extension of trade by the opening of harbours at False Point and Chāndbāli; the district has been covered with a network of canals, which enable the people to transport their goods to the sea and to all parts of the interior; and more recently it has been tapped by the railway and put into direct communication with the mercantile capitals of Calcutta and Madras.

Imports and exports.

The trade of Cuttack, at the present day, consists chiefly in the exchange of agricultural products in their raw state for foreign manufactures, metals and articles of luxury. The chief article of export is rice which is sent to Calcutta, Mauritius and Ceylon, other exports being oil-seeds, hides, jute, timber, horns, feathers, bones, *panasi* grass, and silver filigree work. Jungle products, such as myrobalans, lac, nux vomica, wax, and resin come in from the Garjāts, and with the other articles mentioned above are despatched to Calcutta, and to a smaller extent to the Madras Presidency. English and Bombay piece-goods, kerosine-oil and specie are brought in from Calcutta and sent out again to Purī and the Garjāts along with dried fish and salt; and a portion of the consignments of cocoanuts sent from Purī is forwardd to the Garjāts.

The principal articles of import from Calcutta are English and Bombay piece-goods, kerosine-oil, crockery, glassware, specie, fancy goods, potatoes, lead, copper, zinc, gold, silver and cotton yarn. Most of these articles are consumed in the district, and a portion, as mentioned above, is sent to the Garjāts on the west and to Purī on the south. The articles imported from other places are salt, turmeric, chillies, sugar and tobacco leaves from Ganjām; cocoanuts and cocoanut-oil from Purī; wheat, pulses, oil-seeds, cotton and molasses from the Central Provinces; lac, nux vomica, myrobalans, silk cocoons, wax, resin, iron, horns, hides, timber, catechu, oil-seeds and feathers from the Garjāts. Cotton goods, besides being brought by sea from Calcutta, are also imported by land from some other parts of Bengal, and from certain districts of the Madras Presidency; the imports from Bengal consist of muslins from Midnapore, Bānkurā and other places, and those from Madras of fine cotton goods and silk fabrics sent from Ganjām and Berhampore.

Statistics of trade.

Cuttack is primarily an exporting district, the exports greatly exceeding the imports, whether borne by rail, river, road, canal or sea. The returns of sea-borne trade passing through False Point show that the value of the imports in 1904-05 was Rs. 3,350, while the exports were valued at nearly 22¾ lakhs; nearly all of this was foreign trade, the value of the goods carried by coasting vessels being only Rs. 76,800. The statements showing the volume of rail- and canal-borne traffic show clearly the increasing popularity of transit by rail; and although the canals continue to perform an important function, especially in the export trade, there can be no doubt that as a channel of trade they are being rapidly supplanted by the railway. The latter brought into the district 955,000 maunds and exported over 625,000 maunds of goods, while the canals exported 214,000 maunds and brought in only 31,700 maunds. The export traffic carried along the canals consisted almost entirely of rice, while the cotton twist and yarn brought by boat and steamers from Calcutta constituted nearly seven-eighths of the imports. The returns for rail-borne traffic show that among the imports salt from Madras (404,000 maunds) bulked most largely; a large portion consisted of rice (175,700 maunds) brought in from Purī, which was probably only conveyed into Cuttack on its way to other markets; and after this, the largest import was gram (75,000 maunds) from the Central Provinces. Madras is the largest importer, sending 4⅓ lakhs of maunds, and next come the other districts of Orissa with 205,000 maunds, and then Calcutta and the metropolitan districts with 169,000 maunds. Among exports, rice occupies the first place, nearly 4 lakhs of maunds being exported in 1904-05, and the next most important export was gram and pulse (98,000 maunds); with the exception of paddy, the other exports are small in quantity and of little significance. Practically all the exports are sent to Calcutta and the surrounding districts, or the other districts of Orissa.

The export business in Cuttack is almost entirely monopolized by a few rich merchants, who are in a position to make large advances at favourable rates and to undertake the risk of sending consignments by sea. These merchants have their agents, who establish small depôts in various parts of the district and make large purchases of rice at the markets. They then send out their rice to the chief ports (False Point and Chāndbāli), whence it is shipped to foreign countries or to other parts of India. There are many traders, however, who carry on heavy transactions both in articles of export and of import: they are generally the Telis, Kewats, Guriās, Pātras and Golās of the district, but their numbers are swollen by Bengalis, Mārwāris, Bhojpurīs, Telingas, and Garjāti Jaunliās, who with their greater industry and enterprise have managed to establish themselves to the exclusion of the Oriyās. These dealers (*mahājans*) make advances of money to petty tradesmen, known as *bepāris* or *gumashṭās*, who go to different parts of the country, buying in at the cheaper markets and selling off to advantage at the dearer ones. At the close of the year, or of any other definite period of time, they return home, render an account of their transactions, and pay the *mahājan* the money they have borrowed, together with a share of the profits. The latter is divided between the *mahājan*, the *bepāri* and the owner of the boat or pack-bullocks used in conveying goods from one place to another. Mercantile classes.

Sometimes the *mahājans* have *kothis* or warehouses in different parts of the interior, where they station resident agents or *kothi gumashtās*, who are often near relatives of the *mahājans* themselves. Goods are bought and sold here, and are also sent off to the *mahājan*; at the end of the year accounts are settled, the profits that accrue from the year's transactions being divided among them. Some traders, besides dealing in all kinds of produce themselves, also serve as brokers or middlemen for the sale of miscellaneous goods, and are known as *dandidārs* (literally weighmen), *i.e.*, men through whose intervention purchases and sales are effected.

Many *mahājans* personally keep up wholesale or retail shops, at which they sell their goods either wholesale to peripatetic dealers, or retail to ordinary purchasers. These shops, which are found chiefly in Cuttack, at the sub-divisional head-quarters and at a few other trade centres, are held for the most part by outsiders. In Cuttack town the Mārwāris, who have been settled down there for the last 40 or 50 years, trade chiefly in European piece-goods; the Bhojpurīs, who deal in spices, have been in Cuttack for the last 70 years; the Telingas, who followed the Madras regiments to Cuttack, have established themselves as traders in Deccan clothes, chillies, camphor, etc.; and the Kābulis, as in other parts of Bengal, preside at stalls of dried fruits of all kinds.

Trade routes.

The sea-borne trade of the district is carried by the steamers and sailing vessels which touch at the ports of Chāndbāli and False Point. The former of these is at the mouth of the Dhāmra in the Balasore district, but plays an important part in the trade of this district. The opening of this port about 30 years ago, and of False Point harbour in 1860, at once gave a great impetus to trade and effected quite a revolution in the whole Province. Before the establishment of these two ports, Orissa was an isolated country, but now it has become a busy centre of trade, the sea-borne trade being carried on with ports all along the coast in Bombay, Madras and Bengal, as well as with Ceylon, the Maldive Islands, Mauritius and England.

The land-borne external trade is carried on by land with the Central Provinces, the Garjāts, Purī and Ganjām by means of carts and pack-bullocks, and during the rains by boats, while a certain amount of merchandise is sent along the Orissa Coast Canal. But the railway constitutes the chief artery of commerce; the main trade of the district is that which is carried out of it or brought into it by this route, and the other means of communication by land merely supplement its work.

The internal trade of the district is similarly carried on by the railway, and by carts, pack-bullocks and boats; it consists chiefly of cereals and pulses, country cloths, brass and bell-metal utensils, timber and firewood. Next to the railway the most important routes by land are the Tāldanda road, the Māchgaon road, the Chāndbāli road, the Grand Trunk Road, the Sambalpur road and the Ganjām road.

The routes by water are furnished by the rivers and canals which traverse all parts of the district. The Mahānadī and Brāhmanī, together with their main branches, the Kātjurī, Birūpā and Kharsuā, are navigable in all seasons in their lower courses, though their upper reaches and minor branches generally contain little water during the summer. A steamer service plies on the High Level and Kendrāpāra canals, while the Tāldanda canal is regularly used by boats. These highways of commerce have now ensured a certainty of communication between Cuttack and the sea-board, and have helped most materially to develop its trade. The whole district is moreover intersected by a network of channels and creeks, all of which are navigable by country boats during the rains and some throughout the year. With the district and village roads, they form important factors in opening out the more remote villages, so that there is generally some means of communication between them and the trade marts of the district.

Trade centres. The chief centres of trade are situated along the rivers and canals, or are in the neighbourhood of flourishing villages. The most important markets for rice, the chief article of commerce, are False Point, Jambu, Bhutmundi, Cuttack, Mārsāghāi, Alba, Kenduapatnā, and Bāraboriā, and the most important places of supply are Cuttack, Kendrāpāra, Jājpur, Aul, Patāmundai, and Baideswar. Rice is carried chiefly to Calcutta by the railway or by steamer *viâ* Chāndbāli, the traffic being mostly in the hands of the India General Steam Navigation Company, which maintains a regular service of steamers from Calcutta. A portion of it is also carried to False Point in boats and along the Orissa Coast Canal in steamers.

Markets and fairs. Besides these, there are numerous minor markets or *hāts* all over the district, numbering about one hundred and fifty, which are held every week on certain fixed days for the sale of the ordinary articles of consumption and of every-day use. The most important of these are Ichāpur, Barhāt, Mahīpāl, Thākurhāt, Chāndla, Mahalhāt, Pandālo, Pāilo Antoi, Pasai, Keshpur, Chandanagar, Rājnagar, Kerapur, Ganjā, Olābhar, Talsangā and Nikrai in the Kendrāpāra sub-division; Gandkul, Katrā, Bilhāt, Tānghī, Champāpur, Nischintkoilo, Niāli and Kuhunda in the Sadar subdivision; and Binjhārpur, Kāiparā, Haripur, Katikatā, and Kayāngolā in the Jājpur sub-division. Almost all the trade of the interior may be said to be carried on in these markets. People flock in from great distances to buy stores for the week's consumption, and in many cases the *hāts* form the sole source from which the villagers can obtain the necessaries of life. Shops are very rare in small villages, and are generally of a very primitive kind, unable to supply a single family with provisions even for a week, and totally unequal to meeting the wants of a whole village.

Fairs are also held in different parts of the district in connection with the religious festivals. Some of these last for a day, others for a week, ten days, or even a month. Some of the petty fairs have no influence whatever on trade, but the principal *jātrās*, such as the Paus Sankrānti *jātrā* in *pargana* Kātiā, the Salanuni annual fair in Sargarā, the Mahāvināyaka *melā* in Darpan, the Charak Sankrānti *jātrā* in Madhupur, the Gundichā *jatra* in Tikan, the Asokāshatamī *jātrā* in Jājpur, attract many shop-keepers and merchants from Balasore, Bhadrakh and the Garjāts, as well as from the more important centres of trade in the district itself.

Weights and measures.

Cuttack has a peculiar system of weights and measures. The maund of 40 seers is recognized, but the Balasore seer of 80 *tolās* is used for weighing imported goods, and the Cuttack seer of 105 *tolās* for indigenous goods. These seers are regarded as standard weights and are used in the municipalities and the principal market-places in the interior, though not to the exclusion of other weights. In rural areas, however, the seer and chittack are hardly understood, the weights in common use being the *pal*, *bisā* and *pasuri*, and for grain measure the *gauni*. The *pal* is equivalent to 6 *tolās*, but the *bisā* ranges from 18 to 30 *pals*, *i.e.*, from 108 to 180 *tolās*, according to locality and the commodity weighed. No *bisā* corresponds, therefore, to the Cuttack seer of 105 *tolās*. In some places the people realize the disadvantages of this and calculate the *pal* at 5 instead of 6 *tolās*, so that their *bisā* of 18 *pals* may approximate the Cuttack seer. The *gauni* again varies from 1¾ seer to 7 seers according to locality. It is a basket, and not only does it give wrong measure by losing its shape, but there is a buying and a selling *gauni*, the former being of course the larger; and it can be manipulated during measurement, *e.g.*, by pressing, heaping up and filling loosely. There is thus room for unlimited fraud, and the extent of the evil may be realized when it is remembered that the *gauni* is the universal measure for weighing paddy throughout rural areas. The *pasuri* is equal to 3 *bisās*, and the weight in *tolās* therefore varies according to the weight of the different *bisās* used. In the Jājpur sub-division two special weights are said to be used, viz., the *gar* and *muthā*, which are 24 and 27 *pals* respectively.

Among measures the commonest is the *gauni* mentioned above. Other measures are the *petal* (1 seer 6 chittacks), used for measuring mustard, linseed, etc., the *pāhili* (ordinarily = 1 seer 1 chittack), used for weighing parched grain, etc., the *arhā*, for molasses, which generally = 80 *tolās* in capacity, while country cloth is measured by the *hāth* or cubit, which varies from 18 to 22 inches.

CHAPTER X.

MEANS OF COMMUNICATION.

DEVELOPMENT OF COMMUNICATIONS.

WHEN the British conquered the district in 1803, it was practically isolated from the rest of India, and there was but little internal communication. In spite of the long sea face, few vessels ventured to put in at the surf-beaten coast, and no measures were taken to survey the harbours or ascertain the capabilities of its estuaries. Traffic along the rivers, then as now, was rendered difficult by the enormous volume of water they bring down in the rains and by the fact that in the dry season they dwindle in their upper reaches to small streams running through broad sandy beds. In addition, however, to the natural difficulties of the river route, the vexatious imposts and transit-dues of the Marāthās, as well as the black-mail which they levied, made it impossible for the boatmen to ply their trade with any profit, and these natural channels were practically unused. Throughout the district there was not a road, in the modern sense of the word, in existence. What were then called roads were mere fair-weather cart-tracks, without bridges and without proper ferry arrangements for crossing the numerous water-courses which they intercepted; and the pilgrims to Purī, who are now quickly transported there by the railway, were forced to follow the dangerous route through Nīlgiri and Mayūrbhanj, which in many places passed through dense jungles infested by tigers and other wild animals.

The Orissa Trunk Road.

The first step taken to remedy this state of affairs was the construction of the great Orissa Trunk Road, which was sanctioned in 1811 and completed in 1825. It was not metalled, however, and in 1854 Mr. Ricketts found it in exactly the same state as when he saw it 25 years previously. "It has not improved in any respect," he wrote, "it is the same long heap of mud and clay, always next to impassable for any wheeled vehicles, except for the high-wheeled Cuttack hackeries, and for many months of the year impassable even for them. In some parts, where the soil is sandy, foot-passengers do not suffer much inconvenience at any time of the year; but across the many low plains where the soil is clayey, the difficulty of making any progress in wet weather can hardly be exaggerated. I have myself been nine hours going ten miles."

The condition of Orissa in 1866.

The terrible deficiency of communications which still existed as late as 1866 was made apparent in the great Orissa famine, when "the people were shut in between pathless jungles and impracticable seas, and were like passengers in a ship without provisions." The state of affairs at that time was graphically described by the Famine Commissioners of 1867 as follows:—"The whole province is geographically isolated to an excessive degree. To the north and north-west the hill tracts merge into countries more hilly, wild and inaccessible, by which they are separated effectually from Central and Northern India. There is a precarious traffic with Sambalpur by boats of a peculiar construction, which navigate the difficult river Mahānadī in the rainy season and for a month or two after; for the rest of the year this communication is closed. On the other side, the nature of the coast and the sea is such as effectually to stop all native traffic for the major part of the year. With one exception—False Point—there is no protected anchorage of any kind, and that exception may be said to be in some sense almost a recent discovery. Such being the difficulties on either side of the length of Orissa, the only ordinary mode of communication with the outside world is by the route traversing its length. That, however, is so much intersected by the streams already mentioned, and has been hitherto so little rendered practicable by art, that it is comparatively little used by wheeled carriages; pack-bullocks still predominate at all times; in the rainy season wheeled traffic is quite impracticable; and when the rains are heavy, even pack-bullocks cannot be used. At this day the European officer who cannot obtain a special steamer must find his way into Orissa slowly and tediously, as ancient officers may have travelled in the days of Asoka, and the very post takes several days between Calcutta and Cuttack."

Subsequent measures.

This calamity directed attention to the state of all the Orissa districts, and mesaures were taken to prevent the recurrence of a similar disaster, roads being opened up, the coast surveyed, and canals constructed. The communication with the outside world, which was thus established, effectually broke in upon the isolation of Orissa, and more recently the Bengal-Nāgpur Railway has extended its system through the district. It is now amply provided with means of communication by the railway and an extensive system of roads and canals. The railway passing through it from north to south connects it with Calcutta on the one hand and Madras on the other; the roads place every part within easy reach of the markets; and a network of canals and distributaries covers the whole country.

As early as the 17th century Cuttack was reached by the circuitous route from the Dhāmra river *viâ* Patāmundai, up to which place sloops and brigs were able to come with Government and other stores; and Harispurgarh at the mouth of the Patnā appears to have been a place of resort for coasting vessels. It was here that the English landed when they first advanced from Masulipatam to Orissa in 1633; it then possessed a fair anchorage, boasted a custom-house, and was described as "a place of good strength with whom our merchants hold commerce with correspondency." The harbour is now sand-barred, and even in Ralph Cartwright's time the river could only be ascended in small boats for some 8 miles as far as Kosīda; but the old name of the mouth of the Patuā, Boita-kuliyā or ship-haven, is significant of its former importance. Soon after the occupation of Orissa, the English made attempts to improve and extend the communications by sea. In 1811 a Master-Attendant was appointed at Mānikpatnā in the Purī district, and a Deputy Master-Attendant was posted to Dhāmra. The light-house at False Point was commenced eight years later, and after almost insuperable difficulties had been overcome, the building was finished and the first light lit in 1826. The advantages of False Point as a harbour and port were not discovered, however, till a much later date. During the first half of the 19th century, this anchorage was little known, and almost unused. Although but two days by steamboat from Calcutta, no regular communication existed; no important trade was carried on; and the exports, consisting chiefly of rice, were entirely in the hands of a few native shipmasters from the Madras coast. A traveller landing at False Point found himself as far from Cuttack as if he had never started from Calcutta, while its isolated and jungly situation, and the long, tedious boat-route inland, through dense forests and across malarious swamps, rendered it impracticable for goods or passenger traffic.

WATER COMMUNICATIONS.

Harbours.

About 1862 the newly-started East Indian Irrigation Company perceived its capabilities for the importation of stores, and an enterprising French firm in Calcutta shortly afterwards established an agency for the export of rice. But for several years the arguments against adopting it as a harbour seemed to be irresistible, as it was represented to be a fever-stricken, jungle-buried creek, several days' journey from any large town and with scarcely a practicable channel inland. Colonel Rundall, however, after a careful investigation insisted on its capabilities, and the history of the famine of 1866 proved him to be right. During that year, when Government was anxiously exploring every means of throwing supplies into the Province, False Point harbour formed the main entrance by which food was brought in. The Famine Commissioners reported that it was the best harbour on the whole Indian peninsula between the Hooghly and Bombay, and strongly urged its claims upon Government. The harbour was re-surveyed and deepened, the channels were clearly buoyed off, and it was connected with Cuttack by means of the Kendrāpāra canal. It was confidently expected that False Point would grow into an important harbour and form the entrepôt for the import and export trade of Orissa. A scheme was put forward in 1875 for the improvement of the port at a cost of Rs. 2,33,000; it was suggested that Port Commissioners should be appointed; and the matter proceeded so far that the Bengal Government moved the Government of India to extend the provisions of the Indian Ports Act to False Point. In the next year, however, the Superintendent of Marine Surveys deprecated any large expenditure on the port, and the event has shewn the wisdom of his recommendation. The expectations that False Point would be the great port of Orissa have not been realized; Chāndbāli has become the port for the Calcutta trade, and False Point only serves Orissa for the trade in rice to Mauritius, Colombo and the west coast ports. It is now a port of call for coasting vessels, though large cargo steamers put in at irregular intervals.

Rivers. At first sight it would appear that the great rivers which issue from the western hills and then pour into the sea after traversing Cuttack, should afford a magnificent highway for the products of Central India. The anicuts constructed across them have however cut off direct communication between the lower and upper reaches; during the rains they become dangerous for navigation owing to the high floods they bring down, and during the rest of the year the current is sluggish and the volume of water small. Even in the greatest of these rivers, the Mahānadī, numerous sand-banks obstruct the channel in the dry season, and the boatmen are frequently obliged to dig out narrow channels to allow their craft to pass. In spite, however, of these obstacles, there is uninterrupted communication from above the anicut at Cuttack as far as Sambalpur, though the traffic has been diminishing since the latter district has been opened up by railway. There is some traffic on the upper waters of the Brāhmanī, but the jagged rocks which in places stud the stream render it dangerous for large boats. Both this river and the Baitaranī almost dry up during the hot weather in their upper reaches, and the small depth of water only allows small boats to ply along them. On the other hand, all the rivers have sufficient water to enable boats to pass and repass further down their course, and there is a fair amount of traffic in the lower tidal reaches. With this exception, they are on the whole either too shallow or too uncertain to be very largely used, and they do not possess any great value as trade routes.

Ferries.

The District Board controls 43 ferries, of which the most important is that across the Baitaranī at Chāndbāli. Passengers are charged small fees, and the right of collecting these is sold by auction, all the ferries being leased out annually to the highest bidders.

Canals.

For the reasons mentioned above, but little use is made of the rivers as highways, in spite of the canals which take off from them and provide communication with the sea-coast. The use of boats has, accordingly, been restricted from time immemorial, and the people have always been accustomed to carry the internal traffic of the country along the roads by means of pack-bullocks and carts. They were thus neither sufficiently accustomed to navigation nor prepared to utilize the canals when they were opened; and these waterways have consequently never been used very extensively. The Kendrāpāra and Gobri Extension canals connect Cuttack with the Brāhmanī at Alba, and from there vessels go down the river to Chāndbāli, where the cargo is tran-shipped and carried by steamer to Calcutta. The High Level canal supplies a direct route between the marts of Cuttack and Bhadrakh, and the Tāldanda canal between the Mahānadī and Kātjurī rivers links up Cuttack with False Point. This canal is used by boats, and there is a regular steamer service along the other canals, as steamers carrying passengers and goods ply three days a week between Cuttack and Chāndbāli. The Gobri and Jājpur canals are also important navigable channels; and there are in all 210 miles of canals used for navigation. The total tonnage of the boats using them in 1903-04 was 256,672 tons and the value of their cargoes, excluding rafts, was Rs. 70,79,162.

RAILWAYS.

The volume of traffic on the canals has, however, never been very large, and the trade passing along them has now diminished owing to the competition of the railway. The line of the Bengal-Nāgpur Railway which passes through the district was opened in 1899. It connects Cuttack directly with Madras and Calcutta, and the stations are so placed that they receive all the grain now brought into the head-quarters town and all that is imported from the Tributary States. The finest pieces of engineering work on the line are to be seen near Cuttack; to the north of the town a great bridge has been built over the Mahānadī of girders laid on massive masonry piers, while to the south, where the iron road has been carried over the wide stretches of sand and water extending between this station and Bārang, one of the most difficult pieces of riverine engineering to be seen anywhere in India has been successfully carried out.

ROADS. From the account given above, it will be seen that Cuttack long continued practically roadless even after the British occupation, and that the roads are a creation of the last half century. The district is now very well supplied with these means of internal communication. Some areas are, however, still very difficult of access, such as the *parganas* of Benahar and Khandi and the strip of country between the Tāldanda and Kendrāpāra canals; while in other parts many villages are only accessible by pack-bullocks, and others again can only be reached by boats in the rainy season. This however is only a natural incidence of a deltaic country, where the large rivers intersecting the plains make the construction and maintenance of roads a matter of great difficulty. In spite of these difficulties, there has been a great advance since the great Orissa famine; and there are now 91 miles of Provincial roads maintained by the Public Works Department, 75 miles being metalled, and 16 miles unmetalled, while the District Board maintains 34 miles of metalled and 335 miles of unmetalled roads, besides a number of village tracks with a total length of 445 miles.

Provincial roads. The most important of the Provincial roads is the Orissa Trunk Road, a great highway running from Midnapore to Ganjām in the Madras Presidency. Begun soon after the British occupation, it took the place of the old pilgrim road which had replaced the former route through Nīlgiri and Mayūrbhanj: traces of this pilgrim road are yet visible in ruined but massive bridges, in the Hindu style of architecture, standing in solitude over streams where there is neither road nor traffic. The road was carried as much as possible along the old line and through waste land and jungle; the bridges were built almost entirely of stone taken from the ruined forts and temples in which the Province then abounded; and it was completed as far as Bhadrakh about 1819. It enters the district at Akshuāpadā 46 miles from Cuttack, and, skirting the western hills, it runs along the High Level canal as far as that town, and then divides into two, the Cuttack-Purī road, generally known as the Jagannāth road, and the Cuttack-Ganjām road. It is raised and metalled throughout, and there are ferries across the Baitaranī, Kharsuā, Brāhmanī and Mahānadī rivers.

There are two other Provincial roads, the Cuttack-Sonpur road running along the valley of the Mahānadī through Bānki into the Sonpur State and the Cuttack-Sambalpur road also passing up the Mahānadī valley through Athgarh, which brings down the traffic of the Garjāts.

District Board roads.

The principal District Board roads are:—(1) The Cuttack-Chāndbāli road, 63 miles long, which connects Cuttack with the important port of Chāndbāli. It runs along the Kendrāpāra canal to Kendrāpāra and then strikes north-east to Patāmundai; after this, crossing the Brāhmanī, it goes due north to Aul, and leaving that place continues its north-easterly course across the Kharsuā as far as Chāndbāli. (2) The Cuttack-Tāldanda road, 44 miles long, of which 14 are metalled, running along the southern bank of the Mahānadī, and forming, with the Tāldanda canal, the principal route between Cuttack and False Point. (3) The Kandalpur-Māchgaon road, which takes off from the last road at the 11th mile and reaches Māchgaon after completing a course of 32 miles. Like the Cuttack-Tāldanda road, it is an important trade-route and carries a heavy traffic in grain. (4) The Phulnākhrā-Mādhab road, 25 miles long, which is carried along the border of the district from the 10th mile of the Purī road to Mādhab and then on to Purī.

Besides these, there are a number of cross roads connecting the main roads; and since the construction of the railway the District Board have opened feeder roads wherever they were necessary to connect the stations on the railway with the interior of the district. That body has also taken steps to extend the existing village roads and to add to their number: the task is however difficult, as the villagers are always apt to encroach on them and to divert old roads to suit their convenience. The District Board has done much to improve this state of affairs, but in spite of its activity there are still many places where there are no village roads at all, and often the villagers can only cross the country by wading through creeks infested by alligators.

Bungalows.

The district is very well supplied with staging and inspection bungalows. There are bungalows at every stage (about 10 miles) along all the Provincial roads, the four principal District Board roads mentioned above, and the canals, as well as at Beruan and other central positions.

Country carts.

The cart in universal use is somewhat peculiar in shape. Two poles of *sāl* wood or bamboo about 12 feet long, tied together at one end and about three feet apart at the other, are joined by cross bars at intervals, and this framework rests on a pair of wheels about four feet high and four feet apart. The bullocks are yoked one on each side of the narrow end and will drag half a ton 15 or 20 miles a day on a metalled road. For carrying grain a long coffin-shaped basket of split bamboo called an *odaro*, which will hold about 10 maunds, is fitted on to the cart. In some parts of the district, carts called *khaorahs* are common; these have no wheels, and the framework inclines from the yoke of the bullocks right down to the ground at the back. From the Garjāts come heavy buffalo carts, shorter and broader in make, with low wheels of solid wood; and with the advent of the railway the light little Madras hackeries drawn by a single bullock have become common in Cuttack town.

POSTAL COMMUNICATION.

The district contains 66 post-offices, 15 telegraph offices, and 887 miles of postal communication.

CHAPTER XI.

LAND REVENUE ADMINISTRATION.

THE early Hindu rulers of Orissa recognized no middlemen between them and their subjects, and every cultivator was in theory bound to pay to his sovereign a share, estimated at from one-twelfth to one-fourth, of the gross produce of his land. The nominal proportion was one-sixth at the outside, but in fact was often more. The residents of each village paid their quota through a headman (*padhān*) who, in consideration of the services he rendered in collecting the revenue, was allowed to hold free of all payment a certain share, not exceeding one-twentieth, of the total land under cultivation, and probably also retained some part of his collections as a perquisite of his office. The village accounts were checked by the accountant (*bhoi*), who was also paid by the grant of a few acres free of assessment. These villages were grouped into large divisions (*khand* or *bisi*) of 10 to 50 square miles, the prototype of the modern *pargana*, many of which are still known by their old Hindu names, such as the Nāhākhand and Derābisi *parganas*. Over each of these divisions was an executive officer called *khandpati*, who acted as the representative of the sovereign, and with the assistance of the divisional accountant (*bhoimūl* or *bishayī*) collected the revenue, and handed it on to the head of the district or *desādhipati*. HINDU REVENUE SYSTEM.

This was the system in the regulation provinces of the Hindu kingdom, but along the hill borders and on the scantily-populated littoral the land was held by military chiefs who paid a tribute to their suzerain, and were independent as regards the internal administration of their properties.

The first regular settlement of Orissa was begun in A.D. 1582 by Akbar's victorious general Todar Mal and was concluded in 1591 by Rājā Mān Singh: this settlement is given in Ain-i-Akbarī as 17 lakhs of rupees for the whole of that Province. For the most part the border chieftains were left untouched; but in the central and most highly-cultivated portions of the district a detailed settlement was made, and the rates of rents in every MUHAMMADAN REVENUE SYSTEM.

village were fixed. The hereditary Hindu officials of the *pargana* (*khand* or *bisi*) were confirmed in possession, the *khandpatis* and *bhoimūls* becoming *chaudhris* and *kānungos*, and being entrusted with the collection of revenue and vested with the other rights and liabilities of zamīndārs for the portion of the *pargana* or *taluk* under their direct management. The village headmen were maintained under the appellation of *mukaddam*, an Arabic word meaning headman, their customary right to hold one acre in twenty free of assessment being commuted to a grant of a definite quantity of land as *jāgīr*. Where there were no hereditary headmen or where the *padhān* had been dispossessed, collections were often made through an agent (*kārji*) or farmer (*sarbarāhkār* or *mustājīr*) appointed by the *talukdār*, and many of these developed into hereditary tenure-holders with rights almost equal to those of the *mukaddams*.

The *talukdārs* were remunerated by grants of land (*nānkar*) given for their support, and by deductions of 5 per cent. from the gross collections of the villages in their charge, as well as by permission to collect what they could in the form of octroi, market-dues and other extras. The land-owners got a deduction of 10 per cent., besides one-twentieth on the total collections in the form of rent-free land. Nominally one-twentieth was the customary deduction, but it was probably greater in the case of zamīndārs than in that of *chaudhris*, and was less still for the *kānungos*. For the supervision of these land-holders Rājā Mān Singh grouped the *parganas* into three *sarkārs*, Cuttack, Bhadrakh and Jaleswar, each being placed in charge of a chief executive officer called an *āmil* and of a checking revenue officer or Sadar *kānungo* with several subordinate collecting agents (*gumashtās*), who were paid originally by a percentage on the collections.

The *talukdārs* and superior officers were nominally appointed by the sovereign or his representative, and were so far officials that they acted jointly in making or sanctioning alienations and assignments of the land or its revenues, and could be removed from their office for bad conduct; but under the two centuries of misrule and revolt that followed Akbar's reign, their hold on the land grew stronger, and the right to appoint lapsed into a purely formal custom of confirming the heir of the deceased official; even the *āmil* and the Sadar *kānungo* came to hold large estates, stepping into the shoes of dispossessed *talukdārs*, for whom they stood security. Their office was abolished by the Marāthās, who appointed *āmils* and Sadar *kānungos* of their own, and at the time of the British conquest these ex-officials were found only as holders of large and valuable estates.

Besides the *talukdārs*, there were also a few land-holders in possession of whole *parganas*. They alone were officially styled dependent Hindu *sardārs* or zamīndārs, and were generally either descendants of the old reigning princes, as in the case of the zamīndār of Utikan, who was the Rājā of Kanikā, and the zamīndār of Saibir, who was the Rājā of Patiyā, or were border chiefs (*khandait* or *bhuiyā*), such as those of Darpan and Madhupur. Besides these two classes, a good many superior officials were appointed zamīndārs in return for special services, and it appears that the Marāthās recognized sixteen of these zamīndārs in the Cuttack district. Like the *talukdārs*, they all held under deeds of appointment, and though their position was more honourable, their rights and liabilities did not in any way differ from those of the *pargana* officials.

MARATHA REVENUE SYSTEM.

In 1742 occurred the first invasion of the Marāthās, and from 1751 Orissa became a Marāthā Province under the management of a Sūbahdār. The new conquerors made in theory no change in the fiscal organization, but recognized the people whom they found in possession of the land without asking inconvenient questions. Orissa, the most peaceful part of their dominions, they regarded solely as a source of revenue, and to this end appointed 32 *āmils* to look after the collections, with distraint officers under them, who squeezed the uttermost farthing out of the people. Defaulting *talukdārs* they unhesitatingly dispossessed, and where they found a village headman strong enough to be independent of the *pargana* official, they allowed him to pay his revenue direct into the treasury. To some extent this had been the practice in the later days of the Mughal rule, when numerous independent estates were created by grants to Ministers of State or for the maintenance of the Nawāb's household. These estates consisted generally of numerous small parcels of land in different villages—an arrangement which was inconvenient in many ways, but was perhaps intended to give these officials an interest in touring or to provide them with a foothold in the different villages. The accounts were kept separately as direct collections (*hazurtahsīl*), and were credited to the support of special departments. To the category of petty separate revenue units were subsequently added the alienated estates known as *kharīdāgi*, *i.e.*, purchased estates and the resumed *jāgīrs* of torch-bearers (*masālchi*) or yeomanry (*dograi*); while, as already mentioned, the Marāthās also accepted direct payment from many village headmen. All these estates were known as *mazkuri*, *i.e.*, specified in the rent-roll, and their holders were treated as having exactly the same rights as *talukdārs*, though the position they held was one of less dignity.

During the fifty years of Marāthā rule, the position of the cultivator and payer of revenue (*mālguzār*) was one of extreme difficulty. Farming leases of estates were put up to auction in Nāgpur and were bought sometimes by two or three persons, all of whom came to Orissa and tried to levy what they could, the purchasers, who were known as *āsāmis*, being responsible for the collection of revenue. The Marāthā demands knew no limit, default in payments subjected the *talukdārs* to dispossession, imprisonment and fine, and, when all other means failed, the Marāthā cavalry harried the country and plundered the villages. The *talukdārs*, when pressed, retaliated on the people by levying extraordinary cesses and so-called voluntary contributions, and their exactions were only limited by the fear of driving the tenantry to abandon their villages and leave the land uncultivated. When the Marāthā cavalry appeared, the villagers would fly to the woods driving their cattle before them, only to return when the troops had retired; but sometimes a border zamīndār was strong enough to meet force with force and beat back the invader, or at least compel him to accept reasonable terms.

Along the sea coast, however, and in the mountainous regions on the west known as the Rājwāra, the old organization survived unchanged. This tract included the territories of some of the feudal chieftains who ruled over the barbarous races of the hills and the lands assigned in Todar Mal's and Rājā Mān Singh's settlements to the descendants of the Hindu kings. Many petty Rājās along the coast were reduced to the position of ordinary zamīndārs, but to the last the hill chieftains resisted even the Marāthās with some success. They were periodically plundered, but it was only by a considerable show of force that the Marāthās could compel payment, and amidst the inhospitable wilds and forests the horsemen were at a disadvantage and were as often defeated as victorious.

With all their extortion the Marāthās appear to have collected annually about 11 to 12 lakhs of rupees from the Province, which is less than the estimated revenue of the Mughals; but this was the natural result of the rapacity of the conquerors, which defeated itself by discouraging thrift and throwing large areas out of cultivation.

EARLY ENGLISH ADMINISTRATION.

All that can be definitely said of these early settlements of the Province would appear to be that, while the Mughals made some attempt to proportion the revenue to the assets of the country, the Marāthās wrung out of it all they could, irregularly and ruthlessly, with the result that, when we first approached the question of its settlement in 1803, the Province was found much impoverished and in a state of anarchy. The British Commissioners determined to cancel all balances outstanding from the demands of former years and to base the demand for the current year on the receipts for previous payments after making suitable deductions on account of *abwābs*, excessive assessments and the sums collected in advance by the Marāthās. In 1804 they issued instructions for making the first regular settlement of the Province, which were subsequently embodied in Regulation XII of 1805. It was to be for one year only, 1804-05, and was to be followed by a triennial settlement. One of the most important objects in view was to bring deserted villages into cultivation. As the Marāthās and their predecessors had based their demands on the amount of land actually under the plough, without reference to the amount of cultivable land in each village, very little encouragement had hitherto been held out to cultivators and proprietors to increase the cultivated area. The Oriyā ryot, whose poverty was his only protection against robbery, extortion and oppression, cared only to grow sufficient rice to support himself and his family for the year. Liberal terms, therefore, were to be offered to those who would bring waste lands under cultivation, but the engaging parties were to be bound, in the most positive manner, and under a severe penalty, not to bring or entice ryots from lands already cultivated, but to collect their ryots from without the Company's territories. That such a proviso should be necessary shows the wretched state of the Province under Marāthā oppression and misrule, which forced the home-loving Oriyās to forsake their hereditary fields and take refuge in the wild tracts in the hills.

Early settlements.

This settlement was concluded early in 1805 and was followed by a number of temporary settlements. A triennial settlement was first concluded, and then in 1808-09 another settlement was made for one year, which was afterwards continued for a further period of three years. Other settlements followed in quick succession—in 1812-13, for one year; in 1813-14, for two years; in 1815-16, for one year; in 1816-17, for three years; in 1819-20, for three years; and in 1822-23, for five years. The history of these early settlements is an unfortunate record of assessment on insufficient enquiry and of the enforcement of inelastic rules for the realization of inequitable revenues; many an old Oriyā family was ruined, and the proprietorship of the land fell, in many cases, into the hands of Bengali speculators. In 1804-05 the assessment for the district was Rs. 4,43,000, and the triennial assessment of 1805-08 brought in an increase of two lakhs, which was, however, made with very little reason. The Collector had no information as to the real assets of the estates, for the zamīndārs and *āmils* combined to withhold all papers, and he had to proceed on a very rough estimate of the quantity of land in cultivation and on the reports of interested subordinates. Arrears accumulated rapidly; and in 1806 began the system of putting up defaulting estates for sale in Calcutta, a policy which proved the ruin of many old families and allowed Bengali speculators to buy valuable properties at low prices. By 1816 the demand had risen to over 7 lakhs with but little justification for the increase either in the spread of cultivation or the circumstances of the people. They were disheartened at the constant alterations of the revenue, and many left their estates to be held by the Collector, who in his turn either managed them through *tahsīldārs*, who embezzled as much as they could, or farmed them out to speculators, who rack-rented the ryots. A large portion of the revenue assessed could not be collected, the hardships of our revenue system were aggravated by repeated droughts, and the amount realized fell to 65 per cent. of the demand.

The Khurdā rebellion.

At last, in 1817, the people, driven to desperation by mismanagement, broke out in what is known as the Khurdā rebellion, when the *paiks*, or landed militia, rose in open revolt against the oppression they suffered at the hands of the underlings to whom was entrusted the collection of the revenue and against the tyrannies of a venal police. The rebellion was quickly stamped out, but it served to bring home to the authorities the deep discontent and real grievances of the Oriyās; and in Regulation VII of 1822 Government shortly afterwards proclaimed its intention of concluding a settlement after making a detailed investigation into the circumstances of the Province.

THE SETTLEMENT OF 1837.

The foundations of the present prosperity of Orissa rest on the great settlement of 1835 to 1845. Preparations for this settlement were commenced as early as 1830, and it was held to run from 1837, although the proceedings were not finally completed before 1845. Taught by the mistakes of their predecessors, the Settlement Officers toiled with marked industry and ability to master every difficulty that stood in the way of a fair assessment, and to store up and tabulate the detailed information, as to the material condition and economic circumstances of the country, without which no successful settlement can be made. Their efforts were cordially supported by the authorities in Calcutta, and their reward has been the unquestionable success of the settlement, the implicit confidence of the people in the records prepared by them, and the endurance of their names as household words throughout the districts in which they laboured. The operations cost upwards of 20 lakhs, and the result was an ncrease of revenue of only Rs. 34,980 for all three districts. In Cuttack it was found that the cultivated area dealt with amounted to 697,000 acres, of which 621,000 acres were assessed. The demand was fixed at Rs. 7,14,100, the incidence of revenue being Re. 1-2-5 per acre; and the balance of the assets left to the zamīndārs was Rs. 4,06,900.

The settlement thus concluded was made for 30 years and expired in the year 1867, but the great Orissa famine of the year 1865-66 rendered it inadvisable to undertake resettlement operations when the former settlement was drawing to a close, and that settlement was accordingly prolonged for another thirty years. The history of the rapid recovery of the Province from the horrors of the great famine has subsequently shown that this extreme leniency was scarcely needed, and that a resettlement might well have been made some twenty-five years ago to the advantage of Government and without undue harassment of the people. The result of the excessive prolongation of the former settlement has undoubtedly been the exclusion of Government for a lengthy period from its fair share of the produce of the soil, and the retention by the landlord classes in Orissa during the same period of profits to which they had no equitable right. During the sixty years of the currency of the settlement of 1837, the district developed in every direction in spite of the disaster of 1866; cultivation extended by nearly a third; communications were largely improved, bringing an increase in the volume of trade; and the prices of staple food-crops were trebled, securing largely increased profits to the cultivators.

Settlement of 1897.

The last settlement of the Province was a work of great magnitude; the operations extended over a period of 10 years, from the end of 1889 to the end of 1899, and over an area of 5,000 square miles; rents were settled for a million and a half of tenants, and the Government revenue on nearly six and-a-half thousand estates. In this district the cost of the settlement was Rs. 766 per square mile, the area assessed was 822,500 acres, and the revenue fixed was Rs. 10,99,300, giving an incidence of Re. 1-5-5 per acre. The settled assets were Rs. 20,72,900, and the actual percentage of the assets taken as revenue thus amounts to 53 per cent., or 11 per cent. less than it was at the preceding settlement. The enhancement made in the land revenue was as much as 54 per cent., but even so it was materially less than was anticipated, and a considerable portion of the enhanced revenue was relinquished by the progressive introduction of the new assessment in the case of estates of which the liabilities were largely enhanced.

Among other changes which took place in the course of the last settlement was the multiplication of estates and landlords to an enormous extent, the number of recorded proprietors rising from 5,400 to 31,900. The lands in the direct possession of the landlords increased proportionately, the area (85,800 acres) held by them being a little more than double that existing at the previous settlement and the valuation having risen from Rs. 75,400 to Rs. 2,23,200. This very marked increase in the area and value of the lands held direct by the proprietors was by no means the only benefit obtained by them during the currency of the settlement, as the incidence of *pāhi* rents rose from Re. 1-14-6 to Rs. 2-11-6. The extent to which the landlords were able to absorb the increasing profits of cultivation during the term of the settlement is shown by the wide difference between the financial position of the proprietors and proprietary tenure-holders at the commencement of the last settlement and at the time when the resettlement operations were initiated. At the beginning of the settlement of 1837, the income of the proprietors, after the payment of land revenue, was Rs. 4,06,900, or 30 per cent. of the assets, and of the proprietary tenure-holders Rs. 25,868, or 20 per cent. of the assets. The assets of the proprietors before the resettlement of 1897 were Rs. 18,47,400 and of the proprietary tenure-holders Rs. 2,04,434, whereas the revenue they paid was only Rs. 7,14,100 and Rs. 1,02,741, so that they actually retained 61 and 49 per cent. respectively of the existing assets.

The moderation of the enhancement is shown by the result of the resettlement on the total income of the zamīndārs. This amounted to Rs. 11,33,000 before the settlement and was reduced to Rs. 9,73,600 after it, or in other words by 14 per cent. As already stated, however, the zamīndārs received an income of Rs. 4,06,900 at the conclusion of the previous settlement; so that, while the revenue has been enhanced by 54 per cent., the income of the zamīndārs has increased by 139 per cent. Where the rise in revenue has been large, the enhancement has been graduated over periods extending to 10 years, and even in 1908, when the demand of revenue will be paid in full, the proprietors' income will still be 134 per cent. higher than in 1845, without taking into account any increase by enhancement of rents of non-occupancy ryots or by assessment of any new cultivation which may be added to their profits in the meanwhile. It may reasonably be held, too, that the large expenditure which the State annually incurs in maintaining the canals and flood embankments enhances its claims to land revenue, the canal system alone entailing upon Government a net annual expenditure of about 10 lakhs of rupees, a sum equivalent to nearly half the revised land revenue of the three districts.

PERMANENTLY-SETTLED ESTATES.

The preceding account will show that the revenue system of Orissa differs from that of the rest of Bengal proper, inasmuch as the settlement for the Government land revenue is not of a fixed and permanent character, but is made for a term of years only, subject to an increased assessment at the end of every fresh period. This system obtains in the greater portion of the district, but a certain number of estates are permanently settled. When the English conquered the Province, they found a few landowners of a superior class, descendants of noble families or of high officials to whom large estates had been assigned for their maintenance, subject only to the payment of a quit-rent. Such were the Rājās of Kanikā and Kujang, who were originally members of the royal family of Mayūrbhanj and were established in their possessions about the 13th century A.D., under the Gajapati dynasty, and the Rājās of Aul and Patiyā, who were the descendants of the last Hindu kings of Orissa and were confirmed in their estates by Todar Mal. Besides these, there were a number of estates all along the sea-coast, denominated *garhs* or *kilās*, which were held by chiefs called Khandaits, who, like their more powerful neighbours of Aul, Kujang and Kanikā, paid but a light *peshkash* or quit-rent, kept bands of *paiks*, and were bound to render military service when called upon to do so: few of these, however, had succeeded in maintaining the privilege of paying only a quit-rent and most had by slow degrees been reduced to the position of ordinary zamīndārs. At the time of the British conquest, these petty chiefs made some attempt at resistance, but they gave way on the approach of the troops and were pardoned; and in recognition of their ancient lineage and to secure their loyalty, Government by Regulation XII of 1805 confirmed in perpetuity the revenue tribute or *peshkash* of Aul, Kujang, Kanikā, Bishnupur and Harispur, and secured from enhancement the revenue fixed by the *sanads* granted to the Khandait zamīndārs of Darpan, Sukindā and Madhupur. The latter were adventurers from the north-west, and, though equally independent, were of inferior rank to the chiefs of the sea-coast. They received *sanads* and executed *kabuliyats* for the payment of revenue, while the Rājās of Aul, Kanikā and similar estates executed agreements and received acknowledgments of their right to hold at a quit-rent. There are now 24 permanently-settled estates with a demand of Rs. 99,507.

Kilājāt ESTATES.

Besides these permanently-settled *mahāls* and the temporarily settled estates of the Mughalbandi, there are in Cuttack certain estates, called *kilājāt*, which occupy an anomalous intermediate position, viz., *kilās* Dompāra, Balarāmpur, Ragri, Chausatipāra and Kantājhar. Mān Singh left these estates as part of the Rājwāra, but the Marāthās resumed them, with the exception of Dompāra, which was of the nature of a *jāgīr*. They collected the revenue payable by them direct, but left their former holders in possession; these were all men of ancient Hindu stock, allied to the chiefs of the Garjāt States, who kept up on a small scale the pomp and dignity of independent chiefs. On the British conquest, these 5 *kilās* were brought on the roll of temporarily-settled estates, and their revenue was enhanced on different occasions; Dompāra was, however, restored in 1829 to the position of a permanently-settled zamīndāri, provided only that, in the event of any alienation, the assessment should be open to revision. At the settlement of 1837 the proprietors of these estates claimed a permanent settlement, pointing out that they had similar titles to those of the Garjāt chieftains, and had only submitted to British rule in the hope of obtaining more lenient treatment. It was ruled however that they had no legal right to be treated otherwise than as proprietors of temporarily-settled estates; but as a matter of policy Government allowed the revenue previously paid to continue for the term of the settlement. At the last settlement it was decided to accord these picturesque chieftains specially lenient treatment and to fix the existing revenue of their *kilās* in perpetuity, subject only to revision in the event of the alienation of the whole or any part of their estates.

GROWTH OF REVENUE.

It has already been shewn that the recent settlement has raised the revenue demand of the temporarily-settled estates from Rs. 7,14,100 to Rs. 10,99,300, which is only Rs. 3,88,000 more than in 1809-10, but in a large number of estates the increase was imposed gradually, and the figure given above is the final revenue which will be payable from 1908. In 1904-05 the total current land revenue demand was Rs. 12,19,553, of which Rs. 10,78,226 was payable by 4,705 temporarily-settled estates, Rs. 41,820 by 7 estates held direct by Government, and the balance by 24 permanently-settled estates. The increase in the land revenue now paid is, however, even greater than these figures would appear to indicate, owing to the fact that in the early days of English administration the whole demand could not be collected and remissions had to be frequently given.

Under the rule of the Mughals and Marāthās the persons whom we recognized as proprietors of the soil were, in theory at least, officers of Government, responsible to it for the revenue they collected, and, accordingly, they were not entitled to any remission. But, when droughts or serious floods occurred, the cultivator did not pay his rent, and there is reason to believe that the native rulers recognized such calamities as a valid excuse for short payments, so that the actual collections always fell short of the full demand. When we first conquered the Province, the Bengal Regulations were extended to it, and the assessment, which under the Marāthās had included a considerable margin for remissions and deductions, became a fixed and invariable debt, which the zamīndār had to discharge to the day on pain of losing his estate, in spite of the fact that Cuttack is a district peculiarly liable to suffer from the extremes of drought and flood. The consequences of the attempt to engraft the rigid administration of a permanently-settled Province on a country and people wholly unsuited to it, how one after another the estates of the oldest families of Orissa were sold up and passed into the hands of Bengali adventurers, how even these failed to meet the revenue demand, and collections fell as low as 65 per cent. and the *paiks* rose in rebellion, has already been described. In later years the Government was more liberal; in the 36 years ending in 1866 remissions of the revenue were made eight times on account of floods, and five times on account of drought; and in this way upwards of 22 lakhs, or three years' revenue, was remitted. Since 1866 no remissions of land revenue have been made, though certain sums have been written off in the Government *khās mahāls*. There has in fact been no such necessity for relief, as the resources of the district have increased largely, while the land revenue has remained stationary; and at the same time the construction of canals and the improvement of the embankment system have greatly decreased the liability to loss from the vicissitudes of the season.

REVENUE-FREE LANDS.

On the other hand, the amount which Government derives from land revenue is very much curtailed by the very large number of revenue-free properties. The *pargana* officials of the pre-British period freely exercised the right of gift, and an enormous number of rent-free tenures were thus created. Some of these tenures were resumed at the first regular settlement of the district, but a great many more were confirmed, as it was laid down that all lands which had been held rent-free during the two previous years, 1802-03 and 1803-04, should continue to be so held during the currency of the settlement. They were to be settled with the persons in possession, on their executing agreements to be responsible for the preservation of the peace, and to abstain from the collection of *sair* or other dues of any kind. It was, no doubt, intended that a careful scrutiny should afterwards be made into the validity of all claims to hold land rent-free under the above rule; but, unfortunately, circumstances prevented this investigation being made until long afterwards. The selection of the two years 1802-03 and 1803-04 as those during which the possession of land rent-free gave a *primâ facie* title to the occupier to continue to hold it on the same terms was peculiarly unfortunate, and resulted in a large loss of Government revenue. During those two years, the Marāthās had little leisure to devote to the details of revenue business. Their own superior officers, with no one to supervise them, contented themselves with their own aggrandizement, and did not interfere with their subordinates, so long as the interests of the latter did not clash with their own. The consequence was that every one, from the *āmil* to the *mukaddam*, took advantage of the confusion to appropriate the lands under his charge. Documents, if called for, were easily forged in those days, and the burden of proof that they were non-valid was thrown upon the Collector; and by the year 1808 more than a hundred thousand such documents, affecting at least one-eighteenth of the land in the Province, had been filed in the Collector's office. Many of these claims were known to be fraudulent and invalid, but no attempt was made to sift them till 1837, when a systematic enquiry was begun. Large areas were resumed, but even so 73,252 estates, covering an area of 128,000 acres, were confirmed as revenue-free. These lands were protected in perpetuity from assessment, with the exception of the lands known as *hin hayāti lākhirāj*, which were declared to be liable to resumption on the death of the incumbent; and the duties of the Settlement Department during the last resettlement were accordingly limited to the identification of the lands claimed as *lākhirāj*, *i.e.*, as free of payment of revenue, within one or other of the confirmed grants recorded in the papers of the last settlement, and to recording the amount of land held without title. The area of the land recorded as revenue-free during this settlement was 133,400 acres, or 16 per cent. of the total area assessed to revenue. This is a large area, but it must be remembered that Orissa is the Holy Land of the Hindus and that it supports a very large population of Brāhman priests in attendance at an infinite number of Hindu shrines. The *bāziāftidārs*, or holders of these revenue-free tenures, are for the most part Brāhmans, who look upon themselves as proprietors rather than tenants. They form the village aristocracy and are most undesirable tenants, as they do not generally cultivate the land themselves and are most remiss in the payment of rent, while the sanctity of their caste makes the petty zamīndār unwilling to proceed to extremes and sell up their tenures.

LAND TENURES.

As regards the persons with whom Government made this settlement, some of the real zamīndārs who held proprietary rights under the Mughals or Marāthās had been recognized as Rājās of *Kilās*, or had received permanent settlements of their lands, before the commencement of the settlement proceedings of 1837. The present zamīndārs of Orissa would appear to be descended from all the *talukdārs*, rent-collectors, village headmen, holders of resumed *jāgīrs* and the like, who were found, at the time of the British conquest, to be paying their revenue direct into the Marāthā treasuries, as well as from the holders of the larger revenue-free properties that were resumed and assessed to revenue by us during the early years of the last century. Under these zamīndārs again a class of subordinate proprietors, or proprietary tenure-holders, was recognized at the settlement of 1837, which was composed for the chief part of village headmen, such as *mukaddams*, *sarbarāhkārs* and *pursethis*, or the purchasers or recipients of proprietary rights in small plots of land from the zamīndārs or *mukaddams*, such as *kharidādārs* or *shikmī zamīndārs*.

The above form the proprietary classes, the zamīndārs paying their revenue direct to the treasury, and the sub-proprietors or proprietary tenure-holders paying their revenue through the zamīndārs of the estates within which their lands lay. With them also may be included the *lākhirāj bahāldārs*, or holders of confirmed revenue-free lands, who possess a permanent right to hold their lands free of land revenue, and are independent of the zamīndārs of whose estates their lands originally formed a part, except in so far as they are bound under Act IX (B.C.) of 1880 to pay certain cesses through those zamīndārs.

The tenantry who hold beneath the proprietors or proprietary tenure-holders mentioned above may be divided roughly into seven classes: (1) the *Tankidārs*, or holders of small areas permanently assessed at a quit-rent; (2) the *Nisfi bāziāftidārs*, or holders of resumed rent-free lands assessed for the term of the settlement of 1837 at half rates; (3) the *Kāmil bāziāftidārs*, or holders of resumed rent-free lands assessed at that settlement at full rates; (4) the *Thāni* ryots, or resident cultivators, whose rents were fixed for the term of the settlement; (5) the *Chāndinā* ryots, or holders of homestead lands, whose rents were also fixed for the term of the settlement; (6) the *Pāhi* ryots, or non-resident ryots, who were practically, in most cases, tenants-at-will; and (7) the holders of service and other *jāgīrs* who held their lands rent-free, either in consideration of services to be rendered, or as rewards for services in the past. The above list is not exhaustive, but includes all but a few small classes of the tenants of the district.

Zamīndārs. The average size of the zamīndāri estates in Cuttack (263 acres) is smaller than in any of the districts of Orissa, and the majority of them pay a revenue ranging from Rs. 11 to Rs. 50. The division of proprietary right by inheritance or sale has been carried to great lengths, the average number of recorded proprietors being seven to each estate, and their total number having increased since the last settlement from 5,400 to 31,900. The number of petty estates is greatest in the north of the district, where the tendency is for property to pass into the hands of proprietors drawn from the cultivating classes with infinite subdivision of lands and shares. In the southern *parganas* all the small estates are being gradually absorbed by the *mahājan* families; but in central and eastern Cuttack almost all the big estates are owned by Bengali zamīndārs, and there have been comparatively few changes of proprietorship in the last 50 years. A large amount of property was acquired by Bengali speculators at the revenue sales held at Fort William in the early part of the 19th century, but since the year 1833 very few estates have passed into the hands of aliens, and, on the other hand, while a few of the Bengali landlords have lost their property, several of them have become practically naturalized in Orissa. Their estates are generally managed with justice and fairness to the ryots, though the majority of them are absentees living in Bengal, who rarely pay a personal visit to their property. Seven of the 15 principal estates in the district are held by these Bengalis, but nearly all the small estates are held by Oriyās. The average revenue paid is Rs. 247, and the zamīndārs are on the whole a prosperous class. After deducting the cost of collection and litigation, the minimum profit which reaches the proprietor's pocket is estimated at 70 per cent.; in a vast majority of cases the net collections are not less than 80 per cent. of the demand; and the steady increase of the prices obtained for zamīndāri estates points to the great advantages and the secure return that this form of investment yields to capitalists.

Tenure-holders. Fortunately for the district the chain of middlemen between Government and the cultivator is not very long. The various grades of sub-infeudation inevitable under the Patni Regulations of Bengal do not exist; in most of the estates only the zamīndār intervenes between Government and the ryot; and even where tenure-holders intervene, they are comparatively few. The various grades of tenure-holders appear to have sprung up in the course of the transition of the Oriyā village communities into their modern form, and the prominent feature which distinguishes them from ordinary tenures is that very few of them were created by the zamīndārs during the last century, and that most of them have grown concurrently with the growth of the zamīndāri interest and in spite of the opposition of the zamīndārs. As the superior officers under the Mughal settlement crystallized into *talukdārs*, so the village headmen and accountants tended to become land-holders, and they are now found in the position of proprietary tenure-holders as *padhāns*, *mukaddums*, *pursethis* and *sarbarāhkārs*. The old Hindu name of the village headman was *padhān*, but most of this class were absorbed, under the Arabic form of the name, as *mukaddams* into the Muhammadan system; while the *pursethis* were also headmen whose duty it was to collect the rents, superintend the cultivation and settle ryots in the villages. All these were transformed in course of time into tenure-holders; while the zamīndārs' subordinate rent-collectors, the *sarbarāhkārs*, gradually acquired separate tenures, just as their masters, having been originally rent-collectors of a higher grade, acquired the substantial interest of zamīndārs. Some *sarbarāhkārs* were originally mere servants of the zamīndārs who collected their rents from the cultivators and enjoyed *jāgīrs*; some obtained possession of their villages as farmers only, but gradually obtained a prescriptive right to the tenure as it descended from one generation to another, while others again were *sardār paiks*, who were bound to attend the summons of the chief and paid rent for that part of their village lands not occupied as *jāgīr*. Another important class of tenure-holders consists of the *kharidādārs*, or purchasers of waste lands. It was customary in the days of the Mughals and Marāthās for the superior revenue officers to recognize a species of sale by which those who engaged for the revenue, transferred small areas of waste land and jungle to persons who undertook to bring it under cultivation or to found villages. The areas were supposed to be small and worthless, but fraud was practised in many cases, and valuable lands were frequently alienated for a small consideration.

Shikmī zamīndārs.

The most curious of all the subordinate proprietary tenures is the *shikmī* zamīndāri, which appears to be of recent creation and to have arisen in three ways. The first and perhaps the most common case is that of resumed service tenures, where one or two of the *jāgīrdārs* were permitted to engage for the payment of the revenue, and the others were recorded as *shikmī* (*i.e.* included) zamīndārs paying through the recorded *mālguzārs* or revenue payers. The second case is that in which lands were assigned by a *mālguzār* for the support of his near relatives. The relative, if he did not engage separately for the payment of the revenue, was recorded as a *shikmī* zamīndār and generally received the whole of the *mālikānā* allowance. The third case is that of alienations of one or more villages of an estate previous to the settlement of 1837, when the purchaser was sometimes recorded as a *shikmī* zamīndār getting all the allowances instead of a co-sharer. All these tenure-holders were treated as sub-proprietors at the last settlement. Rents were not settled under the Tenancy Act, but they were allowed to retain a share of the proprietary allowances, and were generally given the whole benefit of any reduction in the proportion of the gross assets given as revenue. They executed *kabuliyats* for the payment of the revenue assessed in much the same form as the zamīndārs, and no attempt was made to curtail or define any of their existing rights and privileges except in so far that they were set forth in the *kabuliyat*.

Alienation of proprietary tenures.

Not only is the market value of these interests very high, but the position and local prestige attaching to the tenures makes them highly prized by the middle-class Oriyās. The tenure-holder is a man of local consequence; the ancient traditions still survive, the *mukaddam* is still the arbitrator in local quarrels, and his authority is held in considerable respect by the tenantry. They were the trusted leaders of the people in Stirling's time, and their position has not changed materially during the last 80 years; they still hold the position of village headmen to whom the police look for help in discovering and detecting crime, and they are practically the interpreters between the rulers and the ruled. The undisputed rights of succession and the freedom of transfer which the *mukaddam* has gradually acquired and which we have finally recognized are however a fatal gift which is gradually undermining the position of these tenure-holders. The devolution of property in Orissa obeys the Mitāksharā law, and the constant splitting up of shares which this system involves has a continual tendency to make the tenure insufficient to support the family. The *mukaddams* therefore either gradually oust the ryots and manage to support their families by paying the rent and cultivating the lands themselves, or they find the struggle hopeless and are forced to sell their ancestral rights. Sometimes they disappear altogether, but more frequently they remain in the village cultivating their lands as the ryots of the new purchasers. All along the Kātjurī river, where the largest number of these tenures are found, the *mukaddams* and *sarbarāhkārs* have been in contact with big and powerful zamīndārs, and in times of bad harvests and scarcity they have been unable to withstand the pressure and have yielded to necessity by selling their interests. Accordingly, a large number of the tenures in the flooded tracts of South Cuttack have passed into the hands of zamīndārs and rich *mahājans;* but on the other hand the village tenure-holders of the protected *parganas* in the interior have for the most part maintained their independence. This dispossession of the middlemen is, however, on the whole a result to be desired. As a landlord, the *mukaddam* collects higher rents than the big zamīndārs; and while he and his relations enjoy complete security of tenure in the best lands, he recognizes no right of occupancy in his other tenants and extorts from them the uttermost farthing. On the other hand, as a ryot he is still strong enough to resist unfair enhancements, and, with the help of the *khatiāns* which he has been given, he may be trusted to enjoy the rights of occupancy the law allows. The number of these sub-proprietors has multiplied greatly owing to the operation of the law of inheritance and partition, and at the last settlement altogether 10,598 tenure-holders were recorded as compared with 1,366 at the settlement of 1837, and the number of tenures recorded was 1,131, of which 593 were *mukaddami*, 281 were *sarbarāhkāri* and 166 were *shikmī zamīndāri.*

Tenants **Thāni ryots.**

The cultivators may be broadly divided into two main groups, the *thāni* and *pāhi* ryots, these being by far the most numerous and important of the tenants. The term *thāni* is a corruption of *sthānī* or *sthānīya, i.e.*, local, and was originally applied to every resident ryot of the village; its use is now restricted to the successors in interest of the resident ryots who were recorded as such in the first regular settlement of the district. The *thāni* ryots have been in enjoyment of a hereditary right of occupancy from time immemorial, and their status is the creature of custom that has been in operation for many generations. Writing in 1822, Mr. Stirling remarked:—"They do not in general take out *pattās* nor do they give *kabuliyats*, as they hold their lands under their *jot* or plough hereditarily, and the amount of their proper payments on account of rent depends on a measurement and adjustment of rates made long anterior to the present day by act of the sovereign." These tenancies, as the holdings of resident tenants, naturally embraced all the best lands of the villages, and the customary rights of the resident tenant included many important privileges. He had the right to take up waste land at privileged rates; he had rights of pasture and fuel; his occupancy was hereditary; his rent was fixed; and he could be disturbed only on failure to pay his rent. The latter was therefore much higher than that of the non-resident tenant; and though it was thought at the settlement of 1837 that it was too high notwithstanding all these advantages, the rents found to be actually paid were as a general rule confirmed. The assessment appears however to have been excessive, and a large proportion of these ryots disappeared during the currency of the settlement, owing no doubt to the famine of 1865-66. The area held by them decreased from 133,700 acres to 83,800 acres and the rents paid by them from Rs. 3,78,700 to Rs. 2,20,900.

Pāhi ryots.

The *pāhi* ryots were originally the non-resident ryots of the villages, who, according to ancient custom, were mere tenants-at-will, until Act X of 1859 and the Bengal Tenancy Act marked an epoch by the creation of occupancy rights for their holdings. But the benevolent intentions of the law and the boon conferred by it were unknown in many parts till the commencement of the last settlement, the word *pāhia* continued to be a term of reproach among the ryots indicative of an absence of rights, and it was only the settlement proceedings of 1897 which brought home to these tenants the nature of their rights. As a matter of fact, the *pāhi* tenant of 60 years ago has ceased to be non-resident. He has settled down on his cultivation, and, under the operation of the Bengal Tenancy Act, he has become an occupancy tenant with all the rights and privileges which that Act has given. The name of the *pāhi* tenant has practically disappeared from the settlement records, as it is now held to comprise all the tenants included under the Bengal Tenancy Act as settled, the holdings of *thāni* ryots and all mixed holdings (*thāni-pāhi*) being excluded. Nearly all of these have occupancy rights, and out of the total area of 427,300 acres, only 8,100 acres are held by non-occupancy ryots.

Under-tenants.

The under-tenants are divided into four classes: (1) ryots of tenure-holders with rights of occupancy; (2) ryots of tenure-holders with non-occupancy rights; (3) under-ryots with rights of occupancy; and (4) under-ryots liable to eviction for failure to pay rent or at the end of the year after service of due notice. From the figures obtained at the last settlement it appears that within the revenue-paying estates there are 120,046 under-tenants holding 52,481 acres. The percentage of under-ryots to the number of ryoti holdings (excluding *bāziāfti*) is 6·9; in other words about 7 per cent. of ryots other than *bāziāftidārs* sublet portions of their holdings, amounting in the aggregate to 2⅓ per cent. of the area held by them, while tenure-holders sublet 44 per cent. of their lands. Nearly all the under-ryots however have other lands of their own held as *pāhi* or *thāni* either in the same or a neighbouring village. Respectable ryots take up and cultivate lands held by Brāhmans, or small patches required for sugarcane and tobacco, which the ryot himself cannot or will not grow; and other under-ryots are tenants paying produce-rents, who are men of low caste with a position differing but little from that of hired servants. Similarly, the ryots holding under tenure-holders may be divided into respectable ryots cultivating for their own convenience and low-caste men paying produce-rents, though the former largely predominate.

RELATIONS OF LANDLORDS AND TENANTS.

The relations between the zamīndārs and their ryots are generally amicable and peaceful, and the partial extension of the Bengal Tenancy Act to Orissa has improved the status of the tenants. There has been an appreciable increase in the number of rent cases under Act X of 1859 during the last two or three years, but this is probably only temporary and is due to the fact that a few large estates which had become disorganized and seriously encumbered have recently been brought under better management; while the last settlement has placed landlords generally in a better position to realize their dues through the Courts. Constant oppression by impoverished petty zamīndārs is certainly less than in Bengal, and there are probably few places in the Province where the relationship between the zamīndārs of all grades and the tenants is more satisfactory than in Cuttack.

CHAPTER XII.

GENERAL ADMINISTRATION.

ADMINISTRATIVE CHARGES AND STAFF.

THE administration of the district is in charge of the Collector under the Commissioner of the Orissa Division, and for general administrative purposes it is divided into three sub-divisions with head-quarters at Cuttack, Jājpur, and Kendrāpāra. The head-quarters sub-division is under the direct supervision of the Collector, while each of the other two sub-divisions is in charge of a Sub-divisional Officer exercising the powers of a Deputy Collector in revenue matters. At Cuttack the Collector is assisted by a staff of Deputy Collectors, which consists generally of five officers, and occasionally of six, when a Joint-Magistrate is posted to the head-quarters station. In addition to this staff, a Sub-Deputy Collector is permanently stationed at Cuttack; another is available for a certain portion of the year to help either of the two Sub-divisional Officers who may be in need of assistance; and there are also two officers employed exclusively on special branches of work, viz., a Special Deputy Collector in charge of excise and income-tax, and a Deputy Collector with certificate powers, who is engaged in the collection of canal water-rates under the supervision of the Superintending Engineer of the Orissa Circle. The Government estate of Bānki, which is included in the head-quarters sub-division, is administered by a Sub-Deputy Collector, who is also in charge of the sub-treasury at that place. The Sub-divisional Officers of Kendrāpāra and Jājpur do not exercise original jurisdiction in any revenue matters except rent suits, all other revenue matters being dealt with by the staff at Cuttack.

After the conquest of Orissa by the British in 1803, two Joint-Commissioners were appointed who at once took measures to place the administration on a satisfactory footing. Courts were established, a land settlement was arranged for, and the Bengal Civil Regulations were extended to the Province. The office of the "Commissioners for settling the affairs of Cuttack" was abolished in 1805, and the Province placed under the charge of a Collector, and of a Judge and Magistrate. For 24 years after this, the whole Province formed but one district, having its head-quarters at Purī until 1816, when Cuttack was made the capital. In 1829 the Province was split up into the three regulation districts of Cuttack, Balasore and Purī, and the non-regulation Tributary States. After the formation of the district, the criminal and revenue jurisdiction underwent many changes until 1870, when the Baitaranī and Dhāmra rivers were fixed as its northern limit. The only important change made in its jurisdiction after that year was the annexation to it of Bānki in 1881. The sub-divisional system was not introduced till 1859, when Jājpur and Kendrāpāra were for the first time constituted separate sub-divisions.

The revenue of the district has gradually increased from Rs. 14,30,000 in 1880-81 (when the income-tax had not been imposed) to Rs. 14,61,000 in 1890-91, and to Rs. 19,71,000 in 1900-01. In 1904-05 it amounted to Rs. 20,40,136, of which the greater portion (Rs. 12,11,384) was derived from land revenue; the other main heads of income were excise (Rs. 3,31,407), stamps (Rs. 3,09,373), cesses (Rs. 1,56,227) and income-tax (Rs. 31,745). REVENUE.

The excise revenue is as usual derived from imported liquors, country spirits, *tāri*, opium, and the duty and license fees on hemp drugs. A statement of the various exciseable articles and of the sums realized from them in the decade 1893—1902 is given in the Appendix, from which it will be seen that the income from this source was steadily progressive till the end of last century. Since the year 1899-1900 it has decreased slightly, and in 1904-05 the total excise revenue amounted to Rs. 3,17,000, which represents an average of about three annas per head. Excise.

More than half of this sum (Rs. 1,90,000) is derived from the sale of opium, a drug to which the people have always been very much addicted. As early as 1813 the Collector reported that the inhabitants of Cuttack might be said to live on opium and that they could hardly exist without it; large quantities were brought in from the Hill States; and when a proclamation was issued declaring all smuggled opium liable to confiscation, the *kachahri* was besieged by applicants for licenses; some *fakīrs* took up their place in front of it with ropes round their necks, vowing that if they were not supplied with the drug, they would hang themselves; and some even offered to pay a tax of Rs. 9 a day for the privilege of buying it. At the present day the use of opium is no less frequent, and the income derived from this source is greater than in any other district in Bengal except Balasore. There is one shop for the sale of the drug and its preparations to every 26,445 persons, and the amount realized from duty and license fees is Rs. 923 for every 10,000 of the people. After opium the most important source of revenue is the duty and license fees levied on *gānja*, *i.e.*, the unimpregnated dried flowering tops of the cultivated female hemp plant (*Cannabis indica*), the amount thus realized being Rs. 70,500 in 1904-05. *Bhāng* is very little used, and the total incidence of the revenue accruing from hemp drugs is only Rs. 346 for every ten thousand, the number of shops licensed to sell by retail being one to every 20,627 persons. The receipts from country fermented liquor or *tāri* (Rs. 40,500) and from country spirits distilled from rice (Rs. 27,550) form the only other source of revenue which is of any importance. The Oriyā is very far from being a hard drinker, and the consumption of liquor is very small. He pays less than half an anna for the luxury of spirits and fermented liquors; the average consumption of distillery liquor per 1,000 of the population is only 3 proof gallons and of outstill liquor less than one gallon, the incidence of taxation per head being only 3 pies and one pie respectively; and the fact that there is only one retail shop to every 82,500 persons shows how small is the demand for country spirit.

The manufacture and sale of country spirit was until recently carried on under what is known as the Central distillery system, *i.e.*, there was one distillery at Cuttack for the supply of the whole district, though this system was modified for the benefit of isolated and sparsely populated areas by allowing a few outstills to be kept up in jungle tracts. Recently however it has been decided to substitute spirit manufactured by European processes for that made locally by the ordinary country methods in Bengal, and to make arrangements, as an experimental measure, with one of the well-equipped distilleries outside the Province to supply Orissa with spirit for sale by retail vendors. The contract distillery system has accordingly been introduced from the 1st April 1905; the local manufacture of country spirit has been absolutely prohibited, and arrangements have been made with the Russa Distillery in Shāhjahānpur for the supply of spirit distilled from rice but manufactured by European processes. The contract suppliers are required to open bonded warehouses and wholesale depôts in places fixed by the Excise Commissioner, so as to ensure the ready supply of country spirit to all retail vendors. The spirit is placed in these bonded depôts and wholesale warehouses at proof or above proof, and is there blended and reduced to certain fixed strengths, at which alone it may be supplied to retail vendors and sold by the latter to consumers. The duty is realized on issue of the spirit from the bonded warehouses, and maximum wholesale prices have been fixed for the period of the contract. The licenses for retail vend are settled by auction, but the contract-suppliers or their agents are not allowed to bid.

The revenue from stamps ranks next in importance as a source of income to that derived from excise, the incidence being a little over 2 annas a head. The people generally are not litigious, and the use of stamped documents is limited, palm-leaf documents being still used for ordinary transactions, though they are gradually going out of use. During the ten years ending in 1904-05 the stamp revenue has risen from Rs. 2,13,000 to Rs. 3,09,000, an increase due to the growing demand for judicial stamps, which brought in Rs. 2,40,500 as compared with Rs. 1,64,400 in 1894-95. The increase in their sale has on the whole been steadily progressive, and has presumably been caused by the gradual growth of litigation, as the receipts from court-fee stamps (Rs. 2,24,000) alone show an increase of over half a lakh of rupees. The demand for non-judicial stamps during the same period has not kept pace with that for judicial stamps, the receipts being only Rs. 20,000 more than in 1894-95. Stamps.

The road and public works cesses are, as usual, levied at the maximum rate of one anna in the rupee. The current demand has recently increased owing to a revaluation which took effect from the beginning of 1903-04; and in 1904-05 it amounted to Rs. 1,71,441, the greater part of which was payable by 6,703 revenue-paying estates, nearly the whole of the remainder being due from 4,228 revenue-free estates. The number of tenures assessed to cesses was 71,046, and there are thus more than six times as many tenures in the district as there are estates; while the number of recorded shareholders of estates and of tenures was 87,666 and 75,546, respectively. The gross rental of the district is now Rs. 34,79,406, as compared with Rs. 21,47,473 when road cess was first assessed under Act X (B.C.) of 1871. Cesses.

From the Appendix it will be observed that in 1901-02 the income-tax yielded altogether Rs. 30,395 paid by 1,347 assessees, of whom 1,021 paying Rs. 11,260 had incomes of Rs. 500 to Rs. 1,000. At that time the minimum income assessable was Rs. 500, but this was raised to Rs. 1,000 per annum in 1903; and the number of assessees consequently fell. In 1904-05 they numbered 502, while the net collections were Rs. 31,745, the incidence of the tax being about half an anna a head. Income-tax.

There are five offices for the registration of assurances under Act III of 1877, one at the head-quarters station and the others at Jagatsinghpur, Jājpur, Kendrāpāra and Tirtol. At Cuttack the Special Sub-Registrar deals, as usual, with the documents presented there, and assists the District Magistrate, who is *ex-officio* Registrar, in supervising the proceedings of the Rural Registration.

Sub-Registrars who are in charge of the other registration offices. The marginal statement shows the number of documents registered and the receipts and expenditure at each office in 1904.

NAME.	Documents registered.	Receipts.	Expenditure.
		Rs.	Rs.
Cuttack ...	6,927	9,493	4,775
Jagatsinghpur...	5,004	3,839	2,216
Jājpur ...	5,744	5,422	2,404
Kendrāpāra ...	4,812	4,000	2,154
Tirtol ...	2,740	1,978	1,575
TOTAL ..	25,227	24,732	13,104

In 1894-95 the total number of documents registered was only 11,943, and even this low figure was 20 per cent. greater than the number registered in the previous two years. From the fact that in each of the three years ending in 1904 the total number of registrations was more than double this figure, it may be concluded that the popularity of registration has steadily grown, though the increase has also no doubt been largely due to the settlement proceedings having stimulated the sale of ryoti rights. The most noticeable features of the various transactions are the number of sales of intermediate tenures and ryoti holdings and the number of perpetual leases registered. Altogether 2,391 deeds of sale of tenures were executed, or nearly double the number registered elsewhere in the Division; 1,427 deeds of sale transferring ryoti holdings at fixed rates were executed, there being only one other transaction of this kind in the rest of Orissa; and there were altogether 8,488 sales of ryoti holdings with rights of occupancy: the purchase-money was Rs. 1,65,534, Rs. 57,835 and Rs. 3,22,166 respectively. The bulk of the purchasers of tenures were tenure-holders, while holdings at fixed rates were generally transferred to ryots, and holdings with rights of occupancy to ryots of zamīndārs. The number of perpetual leases (101) registered was nearly double that of 1903; they were mostly leases of *sebāyat* lands let out by the trustees for large premiums, and it has been suggested that the abnormal increase under this head points to the impoverished condition of the *mārfatdārs* of endowment properties.

ADMINISTRATION OF JUSTICE.

Civil Justice.

The judicial staff entertained for the purposes of civil justice consists of the District Judge, a Subordinate Judge and a Munsif stationed at Cuttack, and of a Munsif at each of the sub-divisional head-quarters, Jājpur and Kendrāpāra. The District Judge is in charge of the administration of civil justice and the other officers work in subordination to him. The Subordinate Judge exercises the usual powers of a Sub-Judge and of a Small Cause Court Judge up to Rs. 500, while the Munsifs exercise Small Cause

Court powers up to Rs. 100. An Additional Munsif is generally stationed at Cuttack to relieve the pressure of work, though he does not belong to the permanent staff, and other Munsifs are also employed temporarily at the sub-divisional head-quarters whenever their services are required to clear off an accumulation of cases. The jurisdiction of the District Judge and Subordinate Judge also extends over the districts of Balasore and Purī. Statistics of the work performed by the various Civil Courts will be found in the Appendix.

Criminal Justice. Criminal justice is administered by the District Judge, who is also Sessions Judge, the District Magistrate and the various Deputy and Sub-Deputy Magistrates at the head-quarters and sub-divisional stations. The sanctioned staff at Cuttack consists, in addition to the District Magistrate, of 4 Deputy Magistrates of the first class and one Deputy Magistrate of the second or third class. Besides these officers, a Sub-Deputy Magistrate with third-class powers is generally posted to the head-quarters station, the Station Staff Officer has the powers of a Magistrate of the third class within cantonment limits, and the Sub-Deputy Collector at Bānki exercises second-class powers. The Sub-divisional Officers of Jājpur and Kendrāpāra are almost invariably Magistrates of the first class, and they are occasionally assisted by Sub-Deputy Magistrates vested with second- or third class powers. In the municipalities of Cuttack, Jājpur and Kendrāpāra there are benches of Honorary Magistrates, exercising second-class powers, composed of 13, 7 and 9 members respectively. One member of each bench is authorized to sit singly for the trial of cases, and one Honorary Magistrate has been granted the powers of a Magistrate of the first class and is authorized to try cases summarily under section 260, Criminal Procedure Code.

Crime. Statistics showing the work of the Criminal Courts and the class of offences dealt with will be found in the Appendix. The latter call for no special comment, as the district is singularly free from serious crime. The Oriyā is generally a mild and inoffensive creature, with little inclination for crime, and he is still further deterred from criminal courses, by the fact that imprisonment often entails loss of caste. Crime is in fact much less frequent than in the more civilized parts of Bengal, nor is there any class of crime characteristic of the district. The commonest forms of offences are theft and burglary, the detection of which is always difficult. The soft mud walls of the houses through which the burglar digs his way renders it unlikely that he will break the weary sleep of the inmates or neighbours, the narrow alleys between the houses afford a safe hiding place, even if a *chaukīdār* happens to pass by, and the adroitness of the thief and the negligence of the village watchman combine to render his trade easy, and his detection a rare occurrence. The property stolen, moreover, usually consists of brass utensils, trumpery ornaments, clothing, cash or grain; and when the same pattern prevails throughout a Province, the identification of the property is as difficult as the concealment of it is easy. The comparative freedom of the district from crime is in striking contrast to the state of affairs in the early part of last century. Cattle-stealing was very prevalent, and large droves of stolen cattle found their way to Calcutta; thefts and robberies were extremely common; but besides these offences, which were probably equally prevalent in other parts of Bengal, the district had an unenviable reputation for the frequent occurrence of cases of murder and homicide. As early as 1815 the Court of Directors called the attention of the Indian Government to the fact that cases of this kind, distinct from and unaccompanied by any other offence, were of more frequent occurrence in Cuttack than in any other district of Bengal. Dacoity, a crime almost unknown before 1810, became frightfully common after the Khurdā rebellion in 1818, though it soon died out with the pacification of the country. Bribery, corruption, peculation, forgery and perjury were rife in all the courts and public offices, though it was but seldom that the offenders were brought to justice. Several cases of *satī* occurred annually, and we learn from the records that in 1812 no less than nine widows sacrificed themselves on the funeral pile of the Rājā of Kanikā.

Police. For police purposes, the district is divided into 10 police circles (thānas), viz., (1) Bānki, (2) Cuttack, (3) Jagatsinghpur, (4) Sālīpur, and (5) Tirtol in the head-quarters sub-division; (6) Aul or Rājābāri, (7) Kendrāpāra, and (8) Patāmundai in the Kendrāpāra sub-division; and (9) Dharmshāla, and (10) Jājpur in the sub-division of that name. Subordinate to the thānas are 14 outposts, and there are therefore, in all, 24 centres for the investigation of crime. The force employed in the prevention and detection of crime consisted in 1904 of the District Superintendent of Police, 4 Inspectors (two in the head-quarters sub-division and one each in the other sub-divisions), 38 Sub-Inspectors, 41 head-constables and 489 constables; and the rural force for the watch and ward of villages in the interior had a strength of 360 *dafadārs* and 3,585 *chaukīdārs* grouped in 737 unions, the average population per union being 2,692 and per *chaukīdār* 553. The cost of maintenance of the regular force was Rs. 71,000, and there was one policeman to every 10⅗ square miles and to every 6,139 persons, as compared with the average of 9⅗ square miles and 4,832 persons for the whole of Orissa. In addition to the regular and rural police, there is a force of 8 officers and 106 men employed in the municipalities and the Cuttack cantonment.

There is a subsidiary jail at each of the sub-divisional stations of Jājpur and Kendrāpāra with accommodation for 12 prisoners each, and a district jail at Cuttack which can hold 340 prisoners. Statistics of the jail population from 1893 to 1902 will be found in the Appendix. The subsidiary jails are merely lock-ups, all but short-term prisoners being sent to the district jail; and in 1904 the daily average number of prisoners was only 6 at Jājpur and 4 at Kendrāpāra. The district jail was built in 1810, and before that time political prisoners of high rank were confined in Fort Bārabāti, while ordinary civil and criminal prisoners were located in huts at Lālbāgh, in the old lines of the European regiments which took part in the conquest of the Province, and in similar buildings near the Magistrate's *kachahri*, which were utilized for the purpose as occasion required. In the early days of jail administration at Cuttack, the labour of the prisoners was employed in the improvement of the town. Those few who were sentenced to private labour remained in the jail and there pounded bricks, made baskets, weaved mats, etc. The rest worked in gangs on the public roads by day, and at night were fastened like a drove of pack-bullocks by chains passing through the rings of their fetters. As a result of this free labour, great improvements were effected in Cuttack, tanks being dug and cleaned, marshes drained, and roads and lanes laid out and repaired. Jails.

At the present day, the prisoners are employed on oil-pressing, rope-making, the preparation of coir fibre, and the weaving of carpets and mats: in 1904 the net profits were Rs. 5,324. Accommodation is provided for 340 prisoners; there are 4 cells for convicts, the hospital holds 31 patients, and there are barracks for 21 under-trial prisoners, 6 civil prisoners, 6 juvenile prisoners, 21 female convicts and 251 male convicts. The average daily number of prisoners confined in 1904 was 312 males and 7 females, and the mortality from all causes was 28 per mille of the average strength. This ratio is higher than that for the Province as a whole (20 per mille), and was largely due to diarrhœa, dysentery and pneumonia.

CHAPTER XIII.

LOCAL SELF-GOVERNMENT.

OUTSIDE municipal areas, local affairs are managed by the District Board which has jurisdiction over the whole district, and by the Local Boards which have been constituted for each subdivision. The District Board is responsible for the maintenance of roads, bridges and road-side rest-houses, and has the general superintendence of primary and middle class schools. It is also entrusted with the management of pounds and public ferries, the control over dispensaries, the provision of a proper water-supply and village sanitation. To the Local Boards, which work in subordination to it, have been delegated the administration of small sums allotted for the construction and repair of village roads and certain functions which will be mentioned later.

THE DISTRICT BOARD.

The District Board consists of 21 members, of whom 10 are elected, 6 are nominated, while 5 are *ex-officio* members. The Statistical Appendix shews, for the years 1892-93 to 1901-02, the principal sources from which it derives its income, and the objects on which it is spent; and it will suffice here to say that its average annual income during the decade ending in 1904 was Rs. 1,34,000, of which Rs. 62,000 was derived from Provincial rates, while the average expenditure was Rs. 1,30,000, of which Rs. 67,000 was spent on civil works, Rs. 44,000 on education and Rs. 6,000 on medical relief. In 1904-05 the Board had an opening balance of Rs. 36,600, and its total income from all sources was Rs. 1,46,000, or 1¼ anna per head of the population. The expenditure in the same year was Rs. 1,60,600. Here, as elsewhere, the Provincial rates form the principal source of income, but the incidence of taxation is extremely light, being only 7 pies per head of the population.

By far the largest portion of the income of the District Board is spent on education, the percentage of the expenditure on education to ordinary income being as high as 40·20. It maintains 5 Middle schools and 3 Upper Primary schools, and aids 18 Middle, 92 Upper Primary, 1,900 Lower Primary and 6 other schools. Besides this, it provides two scholarships of Rs. 20

each a month for education in the Sibpur Engineering College, three scholarships for practical training in the Jobra workshop, and one scholarship of Rs. 12 in the Bengal Veterinary College, and also grants Rs. 100 for the education of women in the Cuttack Medical School. For the purpose of supervision, an inspecting staff of 10 Sub-Inspectors and 30 Inspecting Pandits is entertained.

The District Board maintains altogether 34 miles of metalled roads and 335 miles of unmetalled roads, besides 445 miles of village roads, the cost of maintenance per mile in 1904-05 being Rs. 547, Rs. 55 and Rs. 22 respectively. Special attention has been given to the extension and improvement of railway feeder roads since the construction of the railway, and nearly Rs. 18,000 has been spent in this way during the last two years. In the decade ending in 1903-04, over 16,000 trees have been planted along the principal roads at a cost of Rs. 7,575; there are continuous avenues along 94 miles, and steps are being taken to extend and maintain them according to a definite system. The immediate administration of the roads is vested in the District Engineer, and that officer is also responsible for the management and upkeep of 16 inspection-houses. The Board also controls 93 pounds, from which it derives an income of Rs. 5,500, and 30 ferries, the lease of which brings in about Rs. 11,000 per annum. For the relief of sickness, it maintains 4 dispensaries and aids 3 others, but only 4·4 per cent. of its ordinary income is spent on medical relief and sanitation.

LOCAL BOARDS.

In subordination to the District Board are the Sadar, Jājpur and Kendrāpāra Local Boards, the jurisdiction of each corresponding to the sub-divisional charge of the same name. The system of election in vogue in other parts of Bengal has not been introduced, and the members are nominated by Government. The Local Boards receive allotments from the funds of the District Board, and are entrusted with the maintenance of village roads, the administration of primary education and village sanitation, the upkeep of pounds and the charge of minor works of water-supply. The efficiency of these bodies is reported upon year unfavourably. It is said that, with one or two exceptions the members of the Local Boards render very little assistance in the work of administration, that their attendance at meetings is irregular, and that their supervision and check of village road works and the like is eminently unsatisfactory. Though much useful work is done in the outlying sub-divisions, it is done entirely by the Sub-divisional Officers, and in the head-quarters sub-division, where there is no touring officer to inspect the area in its charge, proper check by the members of the Local Board has been found so impracticable that all roads have been made over to the District Engineer.

MUNICIPALITIES. There are three municipalities in the district, viz., Cuttack, Jājpur and Kendrāpāra. The total number of rate-payers is 11,432 or 15·5 per cent. of the urban population (73,909). The principal source of income in all the municipalities has hitherto been a tax levied, acording to the circumstances and property of the assessees, at the rate of one per cent. per annum on their annual income, but recently a rate levied on holdings has been introduced in the Cuttack Municipality; the total incidence of taxation varies from Re. 1-0-8 at Cuttack to annas 6—9 at Jājpur. Statistics of the annual income and expenditure of each municipality during the ten years 1892-93 to 1901-02 will be found in the Appendix.

Cuttack. Cuttack was constituted a municipality in 1876, and has a municipal board consisting of 18 members, of whom 12 are elected, 4 are nominated and 2 are *ex-officio* members. The area within municipal limits is 4·3 square miles, and the number of rate-payers is 6,895, or 14·8 of the population, the lowest percentage in the Division. The average annual income for the decade ending in 1901-02 was Rs. 48,000 and the direct expenditure was Rs. 42,000, while in 1904-05 they were Rs. 59,000 and Rs. 54,000 respectively. The main heads of income were the tax on persons, from which Rs. 16,930 was realized, and the conservancy rates, which brought in Rs. 15,950, while Rs. 8,300 was derived from tolls and Rs. 4,100 from a tax on vehicles; the incidence of taxation, which is Re. 1-0-8 per head, is heavier than in any municipality in Orissa except Purī. In the same year, Rs. 22,750, or 37·7 per cent. of the total expenditure, was devoted to conservancy, and 17·2 per cent. to public works. Only 5·3 per cent. was spent on medical relief and 2·2 per cent. on education, while the total expenditure on water-supply was Rs. 800. There is no filtered water-supply, and the population have to obtain water for drinking from tanks, wells, and the Mahānadī and Kātjurī rivers. A large proportion of the people are dependent for their supply on the latter river, the scanty flow of which causes considerable hardship in the hot weather. A scheme for the supply of filtered water was prepared some years ago by the Sanitary Engineer, but was abandoned because the cost was beyond the resources of the municipality.

For many years past the finances have been in a very unsatisfactory state, and the municipality has been far from progressive, though Cuttack is the largest town in Orissa and as such should be in a position to act as the pioneer of progress. In 1883-84 the total receipts were Rs. 31,605, in 1893-94 Rs. 39,295, and in 1902-03 they were Rs. 42,651. Of this, Rs. 8,626 was accounted for by the latrine tax, which was first imposed in 1889-90, when Rs. 1,032 was realized, and which brought in Rs. 7,362 in 1893-94. The tax on persons, which is the main source of income, brought in Rs. 15,119 in 1883-84 and Rs. 15,475 in 1902-03, and the receipts under this head were therefore practically the same after 20 years. The increase, such as it was, was mainly due to latrine fees, a tax levied for a special purpose which should not be devoted to other objects; and it was clear that taxation had not kept pace with the expansion of population and the growth of the importance of the town during the past two decades, especially since the opening of the railway. But the most signal instance of financial mismanagement was the fact that the liabilities of the municipality over the assets amounted in 1903 to no less than Rs. 17,655, the deficit representing more than 41 per cent. of the total annual income of 1902-03. For years the municipality had done little to improve the water-supply, to extend and improve drainage, to make new roads, to widen existing ones, and to support educational and medical institutions adequately; in other words it failed to keep abreast of all the principal civic requirements which add to the comfort, convenience, health, and education of the citizens.

In his report for 1902-03 the Commissioner stigmatized the condition of the Cuttack Municipality as lamentable. He stated that, instead of progress, there was a falling back in almost all important respects, and cited their failure to improve the assessment as a discreditable instance of the feebleness of the administration of the Municipal Commissioners of the town. Again in 1904 the Commissioner reported that, though the assessor appointed to revise the assessment increased the tax on persons by Rs. 14,300 and the latrine tax by Rs. 4,400, the total increase of Rs. 18,700 was reduced to Rs. 4,242 owing to the wholesale reductions made by the appeal committees, the net result being most unsatisfactory, apart from the large amount of avoidable harassment and expense which the proceedings entailed upon the people. Even after this increase of the assessment, the percentage of the rate-payers to the total population was nowhere smaller in any of the 16 principal towns in Bengal, and the incidence of assessment was lower than in any other of these towns, except in the much smaller municipality of Balasore.

On this Government pointed out that the Commissioners had only themselves to blame for the nugatory results of the reassessment and for their financial difficulties. It was observed that the fact that a town of the size and importance of Cuttack should be content with the primitive tax on persons was a grave anomaly which ought not to be allowed to continue; and the Commissioners were advised to substitute a rate on holdings for the tax on persons and to undertake a fresh assessment for the purpose. The Commissioners then resolved to undertake a fresh assessment and to introduce a rate on holdings at the rate of 7½ per cent.; an assessor was deputed for the purpose of revising the assessment; and the new rate has been in force since the 1st July, 1905. Strenuous efforts have also been made to effect an improvement in other directions, since the attention of the Commissioners was drawn to the very unsatisfactory condition of the municipality; steps have been taken to ensure the prompt realization of arrears, the careful husbanding of existing resources, the removal of various abuses and the restoration of financial solvency, though it has only been possible to secure financial equilibrium by heavy retrenchments, by severe economy, and by starving the public works. The result of the measures taken may, however, be seen in the increased expenditure on the discharge of the essential duties of a municipality. Taxation has risen to over a rupee a head and the total income to Rs. 58,950, in spite of the heavy falling off of the receipts from ferries since the opening of the railway. The amounts spent on medical relief and roads are now nearly double what they were two years previously, while the expenditure on water-supply has increased five-fold; and it seems probable that with the change in the mode of assessment further developments will be possible. Since the construction of the railway Cuttack has been gaining in reputation as a health resort, its trade has increased, and the letting value of holdings has in consequence been rising steadily. The substitution of a rate on holdings for a tax on persons has therefore been a measure of great importance as a means of raising the tax with the rise in the value of holdings.

Jājpur.

Jājpur was constituted a municipality in 1869 and has a municipal board consisting of 12 Commissioners, of whom one is an *ex-officio* member and the rest are nominated. The area within municipal limits is 2 square miles, and the number of rate-payers is 2,016 or 16·6 per cent. of the population. The average income for the decade ending in 1901-02 was Rs. 5,800 and the expenditure Rs. 5,300. In 1904-05 the income was Rs. 6,885, of which Rs. 4,124 was realized from the tax on persons, the incidence of taxation being as low as annas 6-9 per head. The expenditure in the same year was Rs. 6,465, of which 35 per cent.

was expended on conservancy, 16 per cent. on medical relief and 10 per cent. on education.

Kendrāpāra was constituted a municipality in 1869 and is administered by a municipal board consisting of 12 members, all of whom are nominated. The area within municipal limits is 2½ square miles, and the number of rate-payers is 2,521 or one-sixth of the population. The average income and expenditure for the decade ending in 1901-02 were Rs. 8,000. In 1904-05 the income was Rs. 11,100, of which Rs. 6,662 was derived from the tax on persons, the incidence of taxation being annas 8-9 per head of the population. In the same year the expenditure amounted to Rs. 9,000, of which 21·6 per cent. was spent on public works, 17·6 on medical relief, 15·6 on conservancy and 8 per cent. on education. Kendrāpāra.

CHAPTER XIV.

EDUCATION.

EARLY HISTORY OF EDUCATION.

NOTHING perhaps illustrates the progress of Orissa under British rule more clearly than the history of the spread of education among its people. The contrast between the low estimation in which early observers held their intellectual capacities and the standard which they have now reached is very striking. Orissa was described as the Bœotia of India, and its people as equally ignorant and stupid; it was cited as a proof of the poverty of their qualifications that the principal official posts had to be filled by foreigners; and the reason assigned for this was that it was impossible to find Oriyās of sufficient ability for positions of responsibility and trust. When we first acquired the Province in 1803, there was scarcely a single native of Orissa in Government employ. The language of the courts and public offices was Persian, and it was not till 1805 that orders were passed that in all written communications with the natives of the Province the subject should be written in Oriyā as well as in Persian. This order necessitated the employment of Oriyā *muharrirs*, who, though skilful enough with their iron pen and bundle of palm-leaves, were almost helpless when required to write on paper with an ordinary pen. They are said to have been slow in acquiring any facility in this method of writing, ignorant of business in general, and especially of the new English method of revenue accounts. All the best ministerial appointments were consequently in the hands of Bengali clerks, who, attracted by the high pay that had to be offered to procure the requisite standard of efficiency, left their homes in Bengal, and bringing their families with them, settled in the Province and became naturalized Oriyās. Matters appear to have improved, but slowly, as time went on; and in 1821 the Magistrate reported:—"Scarcely a single real Oriyā receives a salary of more than Rs. 10 per mensem, but several are naturalized Bengalis or Musalmāns. I always give a preference to Oriyās, but at this moment I scarcely know a single Oriyā possessing qualifications to fit him for being a common *muharrir*."

The backwardness of education in Orissa during the first half century of British rule has been graphically described by Sir William Hunter. "Government," he writes, "not less than the missionaries, long found itself baffled by the obstinate orthodoxy of Orissa. Until 1838 no schools worthy of the name existed, except in the two or three little bright spots within the circle of missionary influence. Throughout the length and breadth of the Province, with its population of 2½ million of souls, all was darkness and superstition. Here and there, indeed, a *pandit* taught a few lads Sanskrit in a corner of some rich landholder's mansion; and the larger villages had a sort of hedge-school, where half a dozen boys squatted with the master on the ground, forming the alphabet in the dust, and repeating the multiplication table in a parrot-like sing-song. Any one who could write a sentence or two on a palm leaf passed for a man of letters. In 1838 Government entered the field, and opened an English and a Sanskrit school at Purī. But these institutions proved altogether unable to make head against the tide of ignorance and bigotry, and presently sunk beneath the flood. In 1841 we opened a higher class English school at Cuttack, which after a long series of conflicts and discouragements still survives as the principal seat of education in the Province. During Lord Hardinge's administration two vernacular schools were set agoing in 1845; another one in 1848; and in 1853 an English school was founded in Balasore, while the one at Purī was resuscitated. In 1854 arrived the famous Educational Despatch which was to bring western enlightenment home to the eastern races. Yet for several years afterwards, the increase of schools throughout vast Provinces like Orissa has still to be counted by units. In three great Government estates (Khurdā, Bānki and Angul) we managed between 1855 and 1859 to set on foot 19 elementary schools; but in the latter year the total number for all Orissa, with close on 3 millions of people, amounted to only 29. The truth is the whole population was against us. Such little success as our schools obtained they owed, not to the Oriyās themselves, but to the Bengali families whom our Courts and public offices brought into the Province. Thus, of the 58 Orissa students who up to 1868 reached even the moderate standard exacted by the Calcutta University at its Entrance examinations, only 10 were native Oriyās, while 48 belonged to immigrant families."

The Brāhmans had hitherto held the monopoly of education and kept it strictly in their own hands; and caste prejudice and religious superstition were the great obstacles in the way of progress. The Government schools were looked upon as infidel inventions; and even as late as 1860, a learned Oriyā, on being appointed to the orthodox post of Sanskrit teacher in the Purī school, was excluded for a year or two from the Brahmanical orders, and stormy discussions took place as to whether he should not be formally expelled from his caste. In spite however of such opposition, State education slowly, but surely, made its way in Orissa. In 1848-49 there were but 9 schools, with a total attendance of 279 pupils, out of a population of 3 million souls; but during the next ten years the schools had increased to 29, and the number of pupils to 1,046; while at the close of the third decennial period, *i.e.*, in 1868-69 there were 63 schools with 4,043 pupils.

Until 1869, however, no machinery existed in Orissa for training teachers, and the lack of qualified instructors was one of the greatest difficulties experienced in establishing and maintaining schools. In that year Government opened a Normal school in Cuttack town, at which young men were instructed with the object of qualifying them to become teachers in their turn. On the conclusion of the course of training, these young men dispersed through the Province, and settling in the villages, did much to bring education home to the ignorant peasantry. Each teacher collected as much as he could in money and rice from the villagers who sent their children to his school, and received a small weekly stipend from Government so long as he discharged his duty properly. A considerable number of schools of this sort were gradually opened, and no measure was more successful in breaking down the baneful influences of caste and popularizing education.

PROGRESS OF EDUCATION.

In Cuttack itself the number of schools recognized by Government rose from three in 1856-57 to 50 in 1870-71, and the number of pupils from 168 to 2,755. Between 1871 and 1885 a still more remarkable development took place. In 1872 Sir George Campbell's scheme of educational reform, which extended the grant-in-aid rules to large numbers of hitherto unaided schools, came into effect; and many indigenous institutions being thus absorbed into the departmental system, the number of inspected schools increased by 1875 to 539 with an attendance of 10,196 pupils. The advance of education during the next decade was rapid and sustained, and in 1885 more than 65,000 pupils were receiving instruction in 4,736 public institutions. In other words, the number of schools and scholars was respectively 95 and 24 times as great in 1885 as it was in 1871. This extraordinary rate of progress has not been maintained; in the ten years ending in 1895 the work was hindered by the failure of crops which occurred in several years, the number of schools falling to 3,590, and the attendance to 55,876; and the last ten years have witnessed a similar falling off in the number of schools. These numbered 3,527 on the 31st March 1905, but on the other hand there was a considerable increase in the number of pupils which rose to 65,237; and besides these, there are 203 schools, with 1,870 pupils, which do not conform to any departmental standard and are outside the Education Department system. Thus, during the past decade, the public institutions in the district have decreased by 63, but they have received an accession of no less than 9,361 pupils; and the period has been one of consolidation rather than of expansion. Even so, however, the number of children under instruction is practically the same as in 1885, and it is noteworthy that during these 20 years the scholars studying in Primary schools have increased by only 4,720.

Several causes have contributed to the slow growth of primary instruction. When the Education Department began to devote its attention to the extension and improvement of primary instruction, it had in the first place to deal with a portion of the population which was well-to-do and alive to the value of education, and who lived in the more populous and accessible parts of the district; and it was aided by the existing system of indigenous schools. In such circumstances, progress was comparatively easy. These favourable circumstances have been to a great extent exhausted, and the portion of the problem which remains to be dealt with is far harder. The benefits of education have now to be conveyed to the poorer ryots and the lower castes, who have from time immemorial lived without instruction and are altogether indifferent to it; and besides this, the efforts of the educated classes are more readily directed towards English than towards primary education.

GENERAL STATISTICS.

The contrast between the state of education at the present time and 30 years ago is, however, sufficiently striking; and the wide dissemination at any rate of elementary knowledge among the people is borne out by the census figures which show that whereas the number of literate males was only 57 per 1,000 in 1881, it rose to 109 in 1891 and to 150 in 1901, and that the percentage of literate females rose from 1 to 2 and 5 per 1,000 in the same two decades. Nowhere has the progress in this respect been greater than in Orissa, the number of males over 15 years of age in the whole Division who could read and write being 37 per cent. greater than it was 10 years previously. There seems to be little correspondence between these results and the statistics of persons under instruction compiled from the school returns, as, according to the latter, there was a decrease of nearly 10 per cent. in the number of boys under instruction, while the census shows that the male literate population grew most rapidly. It must, however, be remembered that the comparison between the two sets of statistics is apt to be misleading, as the persons under instruction are for the most part under 15, whereas the estimate based on the census figures deals only with persons over that age, so that a change in the number of pupils in one decade would not have much effect until the following one. There are now altogether 21·6 children at school to every thousand of the population, and there is one school to every square mile or to every two villages. The percentage of boys under instruction to the boys of a school-going age is about 40 per cent.; and in this respect Cuttack occupies a high position among Bengal districts.

The inspecting staff consists of a Deputy Inspector of Schools, 10 Sub-Inspectors and 30 Inspecting Pandits, all of whom are subordinate to the Inspector of Schools, Orissa Division. Statistics of the number of institution and scholars during the years 1892-93 to 1901-02 and of the expenditure on education during 1901-02 will be found in the Appendix.

Collegiate Education. The only college in the district, or indeed in the whole Province of Orissa, is the Ravenshaw College at Cuttack. It was opened as a Zila school in 1841, was constituted a High school in 1868, and was finally raised to the status of a college and affiliated to the University of Calcutta in 1876, when it was given the name of the Ravenshaw College in commemoration of the administration of Mr. Ravenshaw, who was Commissioner of Orissa from 1870 to 1878. A Law department was added in 1871, which was affiliated to the University in 1881; and the College is now composed of three departments—the High school, College and Law department; the number of students on the rolls of the three departments on the 31st March 1905 was 328, 150 and 28 respectively, as compared with 191, 14 and 2 in 1872-73. A Survey school and a Hindu hostel are also attached to the College under the control of the Principal. Instruction is given up to the M.A. standard of the Calcutta University by a staff of lecturers and professors.

Secondary Education.

High English schools. The number of High English schools, *i.e.*, schools teaching up to the Entrance Examination of the Calcutta University, rose from one in 1870-71 to four in 1883-84, and finally to six in 1904-05; during the same three periods the number of scholars attending them increased from 22 to 454 and 1,401 respectively. Of these six schools, four, viz., the Ravenshaw Collegiate school, the Piary Mohan Academy, the Mission High school, and the Town Victoria High school are in the town of Cuttack, and the remaining two are at the head-quarters stations of the outlying sub-divisions of Jājpur and Kendrāpāra. The Ravenshaw Collegiate school, which is, as already mentioned, a department of the Ravenshaw College, is maintained by Government; the Victoria High school is an unaided institution; and the other four are aided by Government under the grant-in-aid rules. The annual cost of the education of each pupil at these schools is Rs. 22-6-2, the cost to the public funds being Rs. 6-9.

Middle English schools.

The number of Middle English schools, *i.e.*, schools teaching up to the Middle Scholarship examination, in which English forms part of the recognized course of studies, increased from 9 in 1872-73 to 18 in 1884-85 and to 27 in 1904-05. One of these schools, which is a practising institution attached to the Cuttack Training school, is maintained by Government, and 4 Middle schools at Charchikā, Mahāsinghpur, Asureswar and Kaliānpur are maintained by the District Board. Of the remaining 22 schools, four are aided by the Education Department and 17 by the District Board, the remaining school being an unaided institution at Ganjā, which is supported by the Kanikā estate.

Middle Vernacular schools.

The third class of secondary schools consists of the Middle Vernacular schools, which read up to the Middle scholarship, but in which the vernacular is the only recognized course of studies. Schools of this class have never been numerous in Cuttack. There were 10 Middle Vernacular schools in 1872-73 and 11 in 1883-84, but they decreased to 5 in 1904-05. Here, as elsewhere, the popularity of these schools appears to be on the wane, owing to parents demanding an English education for their children, and the tendency is to transform the Middle Vernacular into Middle English schools.

PRIMARY EDUCATION.

In 1872-73 there were only 400 children receiving instruction in 17 Primary schools; but the next decade was one of phenomenal growth, the number of these schools increasing to 4,782 in 1883-84, and the number of pupils to 53,013. On the 31st March 1905 there were 3,194 Primary institutions in the district, at which 57,733 pupils were under instruction; of these 55,988 were Hindus, 1,606 were Muhammadans, 59 were Native Christians, and 80 were children of aboriginal descent. The cost of educating each pupil was Rs. 2-11, of which 8 annas represented the share borne by the State. The decrease in the total number of schools which has taken place during the last ten years is to some extent counterbalanced by the increase in the number of children under instruction, but is still very noticeable. Apparently, it is largely due to the disappearance of ephemeral schools under the pressure of competition; small and inefficient institutions have closed their doors, and the pupils have transferred themselves in greatly increasing numbers to larger and more efficient schools. Those which have survived have been given greater stability by the new system of grants-in-aid, under which they receive small quarterly grants supplemented by further allowances at the end of the year; the system of payment by results which was previously in vogue has been abolished, and the payments are dependent on the general condition of the school, as ascertained by inspections *in situ*.

SPECIAL SCHOOLS. The number of special schools increased from 3 in 1870-71 to 28 in 1904-05, and the number of students from 140 to 705; they include all the institutions at which instruction of a special kind is given, such as Training, Medical and Survey schools, Sanskrit *tols* and Muhammadan *madrasas*. The Medical school at Cuttack prepares students for the medical profession, the course lasting 4 years, and the students being trained for the post of hospital assistants in the public service. Candidates for employment as sub-overseers are trained at the Survey school attached to the Ravenshaw College, which it is proposed to raise to the status of an Engineering school. There are in all 5 Training schools. One of these, the Cuttack Training school (formerly the Normal school mentioned above) prepares Head Pandits for Middle schools and subordinate Pandits for Primary schools; three Guru Training schools have been established, one in each sub-division of the district, at which Primary school teachers are trained; and female teachers received instruction at a Training school for mistresses which is attached to the Female Orphanage school at Cuttack.

FEMALE EDUCATION. As in other parts of Bengal, so in Cuttack, female education is still very backward, and the rate of progress has been much slower than in the case of the male population. Considering, however, how intense is the orthodoxy of the Oriyā, the advance has been on the whole very great. In 1870-71 there were 510 girls receiving instruction, and only three schools had been opened; 64 girls' schools have now been established, and the number of pupils has risen to 1,706. Of these schools, two, viz., the Mission Orphanage girls' school and the Ravenshaw Hindu girls' school in Cuttack town, teach up to the Middle Vernacular standard, 5 are Upper Primary schools, and 57 are Lower Primary schools. The Female Orphanage, which is under the management of missionary ladies, is the best conducted of all these insitutions. Besides these, there is a Zanāṇa school at Punang in thāna Jagatsinghpur, which is attended by 26 respectable Hindu ladies.

One notable feature of education in Orissa is the popularity of co-education, *i.e.*, the instruction of children of both sexes in the same school. In Bengal proper the parents have the greatest aversion to boys and girls reading together in school; but here there is scarcely a school in which they are not found in the same class. They study together even in the top classes of Middle schools, and married girls have been known to continue to attend school during the interval between their marriage and the time when they join their husbands.

EUROPEAN EDUCATION. There is a considerable European and Eurasian element in the town of Cuttack, and instruction is given to Protestant children in the Mission European school and to Roman Catholic children in St. Joseph's Convent.

HOSTELS AND BOARDING-HOUSES. Under the rules laid down by Government, students are required to live with their parents or under the care of duly recognized guardians, or in hostels or students' messes. From a special enquiry which was made in March 1905 in order to ascertain how far these rules were observed, it appears that, in the case of the 12 town and 27 mofussil schools of the specified classes, 2,981 out of the 3,569 boys borne on the rolls lived with their parents or guardians, 460 in hostels or boarding-houses, 85 in licensed messes and 43 otherwise, *i.e.*, not in accordance with the rules. In Cuttack town there is a large hostel attached to the Ravenshaw College, at which students of the College, Survey school and Collegiate school can live on payment of 12 annas a month. A hostel belonging to the Training school accommodates all but four of the boys under training, who belong almost without exception to the poorer classes, pay no fees, and are given free quarters. The Mission Orphanage for girls has a large boarding establishment under proper supervision; and small boarding-houses have been established in connection with the Mission and Town Victoria schools. There are also licensed messes attached to the Ravenshaw College, as well as to the school last named and the Piary Mohan Academy; in the case however of the latter two schools they are extremely insignificant. In the mofussil 18 out of the 27 schools have hostels or boarding-houses, and only two schools have licensed messes.

The conditions of the students' lives in the boarding-houses and messes are far from satisfactory in many ways; there is practically no supervision of the boys out of school hours, and the surroundings are often insanitary. The general state of affairs may be gathered from the account of one mess which was visited; this mess consisted of 10 boys paying Re. 1 each as house-rent and Rs. 5 or Rs. 6 each for messing; the courtyard was malodorous and insanitary, and the boys had only one small kerosine-oil lamp between them by which to read; this lamp cost 6 annas and must have been ruinous to their eyesight. In the mofussil the boarding-houses are generally built in the school compound, and one of the masters is nominally in charge of the boys, but he does not always remain on the spot, and the boys are not properly looked after. Parents, it is said, not infrequently bring their boys from distant places and leave them at the school with a certain quantity of rice, etc., to carry them through term-time, and then expect them to fend for themselves; and it is reported that many of the houses set apart for hostels are more fit for the accommodation of cattle than of boys.

CHAPTER XV.

GAZETTEER.

Alamgīr Hill.—A peak of the Assia range of hills in the Jājpur sub-division, situated in 20° 37′ N. and 86° 14′ E., and rising about 2,500 feet above the level of the surrounding country. On the summit of a precipice overlooking the stream of the Birūpā stands the mosque of Takht-i-Sulaimān, the white walls of which form a conspicuous mark on the hill side visible for many miles to the south. It is a plain building, consisting of a single room, surmounted by a dome, and bearing an inscription in Persian, engraved on three seals of black chlorite which form the frieze, denoting that the building was erected in 1132 A.H. (1719-20 A.D.) by Shujā-ud-dīn, the Orissa Deputy of the Nawāb Murshid Kulī Khān.

The tradition connected with the building of the mosque runs as follows:—On one occasion, the Prophet Muhammad was winging his way in mid-air on his celestial throne, accompanied by a large retinue. When the hour for prayer arrived, he alighted on Naltigiri. But the throne being too heavy for the hill, and the hill too small for the retinue, the latter commenced to shake and sink. The Prophet became annoyed, pronounced a curse upon it, and repaired to the precipitous rock upon which the mosque now stands. There he offered his prayers, and the print of his knees and fingers is pointed out on a stone which is preserved in the shrine. His followers rested on the four peaks. No water being obtainable on the hill, the Prophet struck the rock with his wand, and a bubbling spring of pure water at once rose up. Tradition also relates that when Shujā-ud-dīn was marching to Cuttack, he encamped at Irakpur, where he heard the voice of prayer chanted from the top of the hill at the distance of six miles. His followers became anxious to visit the shrine, but Shujā dissuaded them, making a vow at the same time that, should his march prove successful, he would come back and pray on the spot with them. On his victorious return, Shujā constructed a road up the hill about two miles in length, and built the mosque which still bears his inscription.

The ascent is from the east and consists of a steep road paved with rough stones, which still retain some semblance of steps. In front there is a platform surrounded by a thick wall with a gate. Towards the west, high rugged peaks overlook the building; on the north, a high terrace has been raised for the reception of dervishes and pilgrims. On the southern side of the mosque, on the edge of the precipice, is the sacred tank, a small shallow hole cut in the rock, about 10 feet by 8, and 3 feet deep. It is now dry, but the legend is that it was formerly a spring of water formed by Sulaimān striking the rock with his staff. The tank was said to have been full of water till Shujā-ud-dīn's time, when a soldier of his army having outraged a female pilgrim to the shrine, the spring dried up and has never flowed since. The soldier and the woman were buried at the foot of the hill, and every passer-by throws a stone on the grave, which has thus become a huge cairn by the road side. The expense of the shrine is covered by the profits of an endowment of sixty acres of land granted by Shujā-ud-dīn. The mosque is lighted every evening, and the rocks resound with the voice of prayer every morning and evening, when the people of the neighbourhood, Hindus as well as Muhammadans, offer homage at the shrine. The hill on which this mosque stands is called by the Hindus Baradihi or the great site. The old Hindu name of the Alamgīr peak was Māndaka, from the village of that name at its foot, where the *māndu* or primitive ordeals by means of fire, boiling oil, etc., were held in the ancient Hindu period.

Alti Hills.—A name sometimes applied to the Assia hills owing to the fact that many of the peaks lie in *pargana* Alti. *See* Assia Hills.

Amrāvati Hill.—A hill in the Assia range, which is now known as the Chatiā hill, from its proximity to the village of that name on the Cuttack Trunk Road. *See* Chatiā Hill.

Assia Hills.—A range of hills in the Jājpur sub-division, lying between 20° 35′ N. and 86° 14′ E. None of the hills are of any great height, the highest not exceeding 2,500 feet in elevation, but they are of great interest on account of the sanctity of the shrines which crown their summits and the ruins of ancient temples, forts, sculptures, etc., which they contain. The ancient Hindu name for these hills was Chatush-pitha, subsequently corrupted into Chār-puli, or the four seats or shrines, a name derived from the four highest peaks of the chain, the Alamgīr hill mentioned above, the Udayagiri hill, the Baradihi hill and the Naltigiri hill.

Aul.—One of the six great *kilās* of Orissa, the proprietors of which were granted the right of paying a quit-rent, exempted from enhancement, by Regulation XII of 1805. This *kilā* covers an aggregate area of 139 square miles and comprises the 3 *parganas* of Derābisi, Utihār and Kutubshāhi. It was granted in the reign of Akbar to a descendant of the Hindu sovereign Telinga Mukunda Deva, and has continued up to the present day in the possession of his heirs. At the time of the British conquest the estate was held by Rājā Rām Krishna Deva, with whom it was settled on a permanent annual quit-rent or of *peshkash*. His son, Pratāparudra Deva, having impaired his mental faculties by dissolute habits, became incapable of managing the estate, which was in consequence taken under the management of the Court of Wards. The estate remained under the management of the Court till 1847, when Rājā Pratāprudra's son, Padmanābh Deva, having attained his majority, took charge of the estate with an accumulated treasure of Rs. 85,000. The young Rājā, however, soon got into the ways of his father, and so heavily encumbered the estate with debts that it remained under the attachment and administration of the Civil Court for sixteen years, from 1868 to 1883, when it was released.

Part of the area comprised in the *kilā* was surveyed during the settlement of the district between 1889 and 1899, and the remainder was surveyed and settled in 1900-01. It was ascertained that of the total area 82 per cent. was under cultivation, 10 per cent. being twice cropped, and that rice was grown on 77 per cent. of the cropped area. Rents were settled for 25,387 out of 64,497 tenants, and their rental was increased from Rs. 1,00,363 to Rs. 1,07,533, or by 7·1 per cent. The increase was obtained by the assessment of invalid rent-free tenancies and of excess lands in the holdings. The present rates of rent are Rs. 2-7 and Rs. 2-11-11 per acre.

Bānki.—A large Government estate, covering an area of 117 square miles, and surrounded by the Tributary Mahāls of Orissa, among which it was once included. It is bounded on the north by the Tributary States of Barāmbā, Tigiriā and Athgarh, and on the south by the Government estate of Khurdā; its eastern boundary is the estate of Dompāra, while the Tributary State of Khandparā lies to the west. There are few hills in Bānki itself, but it is surrounded by the hilly ranges of the Tributary States, the outline of which forms a picturesque back-ground and presents some magnificent scenery. The Mahānadī passes through it from west to east, and the greater part of the estate lies low and is submerged in high floods to a considerable depth. Formerly there were no embankments on the Mahānadī, but only jungle extending along both its banks; and it was not till Government took over the management of the estate that the jungle was cut and embankments were constructed. These protective works, however, have gradually broken and been abandoned, with the result that the low-lying country is exposed to flood, and some lands which were formerly cultivated have been covered with sand and thrown into waste. The country is generally open, and there is practically no forest, except for a narrow strip of *sāl* about five miles long, which stretches along the Khurdā boundary to the south.

Until 1839 Bānki was a Tributary State, the property of the Rājā of Bānki. In that year the Rājā, having been convicted of murder and sentenced to imprisonment for life, was dethroned, and his territory was confiscated by Government. From 1839 to 1882 Bānki was under the management of the Superintendent of the Tributary Mahāls, but in 1882 it was annexed to the Cuttack district, and it has since been treated as a Government estate. It is in charge of a Sub-Deputy Collector, who is vested with second-class magisterial powers and with those of a Deputy Collector for the trial of rent suits. It contains a sub-treasury, and for the purposes of administration it is practically, though not formally, a sub-division.

The estate was settled in 1844, after measurement, for 10 years, and again, after measurement, for 14 years from 1854, but the term of this settlement was extended to 1888 in consequence of the Orissa famine of 1866. The last settlement was begun in 1888 and completed in 1891; it expired in September 1905 and resettlement operations are now in progress. This settlement having been made under Act VIII (B.C.) of 1879, a new record-of-rights has to be prepared and a settlement of rents made in conformity with the provisions of the Bengal Tenancy Act. The cost of resettlement has been estimated at Rs. 14,145, as against Rs. 38,613, the cost of the last settlement, and the proceedings are expected to yield a net increase of Rs. 5,816 on the revenue which Government derives from the estate.

As the result of the last settlement proceedings, the gross rental was raised from Rs. 23,866 to Rs. 33,107 and the net revenue from Rs. 21,421 to Rs. 28,820, or by 34 per cent. in both cases, the increase being mainly due to the addition of 4,210 acres to the cultivated area. The latter increased from 33,000 acres to 37,210 acres during the currency of the previous settlement, and was therefore about 50 per cent. of the total area of the estate (74,530 acres), while 11,840 acres were returned as culturable waste. There can be no doubt that this settlement was lenient and favourable to the cultivators, as the average selling price of ryoti holdings rose from Rs. 11-4 in 1863-64 to Rs. 88-4 per acre in 1886-87, or by 800 per cent., the price of food-grains rose by 100 per cent. in the 50 years ending in 1891, and the area under cultivation increased by nearly 13 per cent.

A very noticeable feature of the management of the estate is the work done by the *sarbarāhkārs*. In Bānki they hold the same position as in Khurdā, *i.e.*, they are farmers and public accountants, and are responsible for the total demand due to Government, whether they collect it from the ryots or not. They have no rights beyond those conferred on them by their engagements, and are liable to dismissal by the Collector for misconduct. In making appointments to the post of *sarbarāhkār*, a relative of a deceased or retired *sarbarāhkār*, who has rendered good service, is preferred, provided he is otherwise qualified. They receive a commission varying from 10 to 20 per cent. of the demand, and are also allowed to take the profits of new cultivation for the term of the settlement. Besides this, a remuneration of 5 per cent. on the collections is granted every year to each *sarbarāhkār* who is found to have kept the settlement records in proper order and up to date.

Bārabāti.—The old name of the fort at Cuttack. *See* Cuttack town.

Baradihi Hill.—The highest of the four chief peaks of the Assia range situated about 16 miles to the south-west of Jājpur. The old chieftain of the *kilā* had his seat at the foot of the hill, and the remains of a fort may still be seen; but though the main gate is still standing, the building is in ruins and overgrown with jungle.

Chāteswar.—A village about 12 miles to the north-east of Cuttack, in the Sālīpur thāna of the head-quarters sub-division, containing a temple of Siva, in the porch of which is a stone slab with an inscription in the Kuthila character stating that the temple was built by order of the king, Ananga Bhīma Deva (1119—21 A.D.).

Chatiā Hill.—A hill in the Jājpur sub-division, situated in 20° 37′ N. and 86° 34′ E., near the village of the same name on the Cuttack-Balasore road. On the east side of the hill are the ruins of a fort, called Amrāvati, which is rectangular in shape, with massive walls of laterite and one gate facing east. Within the ramparts is a high platform accessible by a flight of steps, which marks the site of the old zanāna rooms, but a number of broken pillars and capitals alone remain to show the proportions of the building which once stood there. On a smaller platform stood a temple now fallen, and the only remains of the edifice are the images of Indra and his wife Indrānī, life-sized figures cut on solid blocks of slate-stone and carved with some taste. According to local tradition, Amrāvati fort was one of the five *katakas* or citadels of the old Hindu kings of Orissa and covered an area of two square miles; it is said that the great wall which surrounded it was demolished by the Public Works Department for the sake of the stone, which was used for the construction of the Orissa Trunk Road. On the western side of the hill is a small cave with a verandah in front, which is probably the work of Jaina ascetics. The cave is without ornamentation, and has never been thoroughly explored.

Chaudwār.—A village on the north bank of the Birūpā river, opposite the town of Cuttack. It contains the ruins of an ancient fort, the walls of which are still traceable. Chaudwār is believed to have been for a long time one of the chief seats of the power of the old Hindu kings of Orissa; a copper-plate grant of the 6th or 7th century A.D. was dug up here some years ago; and tradition asserts that the walls of the fort were 2 miles long on each side. The enclosure still contains numerous mounds and several temples.

Cuttack sub-division.—Head-quarters sub-division of the district, Bengal, lying between 20° 2′ and 20° 42′ N., and 85° 20′ and 86° 44′ E., and extending over 1,562 square miles. Its population was 1,035,275 in 1901 against 981,991 in 1891. The west of the sub-division lies on the fringe of the Chotā Nāgpur plateau, while on the east it is bounded by the Bay of Bengal. The central tract is a fertile and densely populated plain intersected by the Mahānadī and its offshoots. The density for the whole sub-division is 663 persons to the square mile. It contains one town, Cuttack (population 51,364), its head-quarters, and 2,599 villages. The area under cultivation in 1903-04 was returned at 805 square miles, of which 170 were irrigated from the canals, and the area of culturable waste at 62 square miles.

Cuttack town.—The capital of the Province of Orissa, and the administrative head-quarters of the district, situated 253 miles from Calcutta in 20° 29′ N., and 85° 52′ E. The town stands nearly at the apex of a triangle, the two sides of which are formed by the river Mahānadī and its branch, the Kātjurī. It is a trade centre of some importance and is well provided with means of communication. The Orissa Trunk Road passes through it, and the principal roads in the district converge on it; besides this, it is served by the Mahānadī and is connected by canal with Chāndbāli and False Point. There is also a railway station of the Bengal-

Nāgpur Railway at Chāuliāganj on the outskirts of the town. Cuttack is not only the head-quarters of the district but also of the civil Division of Orissa, of the Orissa Circle of the Public Works Department and of the Orissa Division of the Education Department; and as such, it contains the offices of the Commissioner, the Superintending Engineer, the Executive Engineers in charge of the three sub-divisions of the Orissa Circle, and of the Inspector of Schools. Besides these offices, those of the district staff, the jail, and the various courts, the chief public works are the stone embankments by which the town is protected from inundation, the railway bridge across the Mahānadī, the great anicut on that river which feeds the canals, and the canal workshops at Jobra. The town also contains a General Hospital, lunatic asylum, three churches, one for Roman Catholics, another belonging to the Church of England and a third to the Baptist Mission, a convent and several educational institutions, of which the most important is the Ravenshaw College. The population, which was 42,667 in 1872 and 42,656 in 1881, increased to 47,186 in 1891 and to 51,364 in 1901, including 4,810 persons in cantonments. Of the total number of inhabitants, 40,320 are Hindus, 8,886 are Muhammadans and 2,047 are Christians, and there are also a few Jainas and members of the Brahmo Samāj.

The town itself extends from the Mahānadī on the north to the Kātjurī on the south, and covers a large area amounting to about 4 square miles. It is practically divided into two parts:—the native town straggling along the bank of the Kātjurī and extending northwards to the cantonment boundary, and the cantonments* which run along the southern bank of the Mahānadī. The houses of the Europeans extend along both sides of the road running parallel to the latter river, and those on the northern side of the road command a magnificent view over the broad waters of the Mahānadī with a long chain of wooded hills lining the horizon. On the southern side of the town are the Commissioner's and Collector's offices, built on a huge stone embankment, which protects the native quarter from the great floods of the Kātjurī; the view across its wide bed, set off by the undulating hills to the south-west, is scarcely less attractive than that across the Mahānadī.

The picturesque appearance of Cuttack attracted the notice even of the staid Muhammadan historians, and the author of the Sair-ul-Mutākharin gives the following description of it:—"The

* It has been decided to withdraw the troops from Cuttack and to give up that place as a military station.

ground wherein the fortress and the city of Cuttack are seated is an island surrounded by the waters of the Mahānadī and those of the Kātjurī. The parts that are washed by the two rivers are surrounded by a strong wall with squared stone serving as a dyke or mound against their inundations. For those rivers which are fordable for one half of the year swell so much in the rains that the Mahānadī becomes a mighty stream of about 2 *koo* (or 5 miles in breadth), and the Kātjurī of half as much. The fortress of Bārabāti is seated on the Mahānadī, and is about 3 *kos* in circuit; it is built of stone, brick and mortar with a great deal of art. But the city of Cuttack itself stretches on the lesser river at about 2 *kos* from the citadel. The Governor's palace and the houses of the nobility and principal citizens (which in general affect the waterside, and are mostly seated on the said mound) rising by five and ten yards above the mound, cut a handsome appearance; and they overlook on both sides of the water a fine extensive plain that stretches from 4 to 5 *kos* around. The horizon is bounded by a forest of beautiful, lofty trees, that extend as far as the eye can reach, and line the bottom and sides of a chain of high mountains that seem to reach the very sky; and this beautiful prospect, with its triple circle of beauties, is enjoyed by the inhabitants the whole year round."

History of the town.

According to the legendary account preserved in the Mādalā Panjikā, or palm-leaf records of the temple of Jagannāth, the founder of Cuttack was Makar Kesarī, a warlike prince who reigned from 953 to 961 A.D. Perceiving the military strength of the tongue of land where the Mahānadī first divides into its several branches, he is said to have established a town on the strip of land between the two rivers, and to have protected it from inundation by means of a masonry embankment several miles long. The same chronicles state that Matsya Kesarī, a monarch who reigned in the middle of the 11th century, strengthened the new capital by an outlying fortress on the southern bank of the river, and thus commanded the various channels into which the Mahānadī, the highway between the hills and the plains, bifurcates. Mr. Stirling gives practically the same account. After explaining that the etymology of the word Cuttack is *Katak*, signifying in Sanskrit a royal residence or seat of empire, and that it was distinguished from four other *Kataks* by the designation Birānasī or Benares, he states that it became a capital city as early as the end of the 10th century, during the reign of the Kesarī princes, and that Chaudwār, Jājpur and Piplī divided with it at different periods the honour and advantage of accommodating the Hindu Court of Orissa. The account of the foundation of Cuttack by the Kesarī kings cannot however be regarded as authoritative, as the chronicles of those kings given in the Mādalā Panjikā are believed by many scholars to be unreliable; but there can be no doubt that Cuttack was the capital of the indigenous kings of Orissa from a very early date. For this it was admirably adapted by the natural strength of its position which rendered it a safe place of defence. To quote the Sair-ul-Mutākharin:—"As this spot of fortunate ground is surrounded on every side by the waters of two rivers, such a situation renders it very strong; and should any enemy attempt to besiege the place by coming to an understanding with the neighbouring zamīndārs, and the siege should chance to be protracted until the beginning of the rainy season, he would find it difficult to subsist, and his convoys would be greatly at a loss how to approach his camp. But independently of that, the country round this island, and indeed throughout the whole of Orissa, is very difficult ground, especially about the rainy season, when it becomes so very intersected by frequent rivers and endless deep torrents, that an enemy would find it impossible to reach the end of his journey." The natural strength of Cuttack was still further increased by Mukunda Deva, the last Hindu king of Orissa, who built the great fort of Bārabāti on the southern bank of the Mahānadī.

On the subjugation of Orissa by Kālā Pāhār, the Afghān General of Sulaimān Karānī, the fort passed into the hands of the conquerors, who did not however remain long in possession. In 1575 A.D. Dāūd Khān, the last Afghān king of Orissa, was defeated by Todar Mal and Munim Khān at Mughalmārī, and taking refuge in Cuttack executed a treaty there, by which he was allowed to retain Orissa on ceding Bihār and Bengal to the Emperor Akbar; but in 1576 his disastrous defeat and death at Rājmahāl left the way clear for the Imperial forces, and Cuttack became the capital of the Mughal Sūbahdārs. In the troubled times which followed during the viceroyalty of Alī Vardi Khān, it again became the centre of fierce conflicts. Alī Vardi Khān first had to wrest it from the grasp of Murshid Kulī Khān, the brother-in-law of his predecessor, and then, when the people rose in revolt against the oppressions of his Deputy, he was forced to march again to Cuttack with an army of 20,000 men. Mirzā Bakr Alī Khān, who had assumed the Government, was encamped with his troops and artillery on the southern bank of the Mahānadī, but Alī Vardi Khān's soldiers, plunging into the river, quickly crossed to Cuttack at the Jobra *ghāt*, and dispersing the opposing forces entered the town in triumph (1741 A.D.). The

Marāthās now, however, began to overrun Orissa, and for the next ten years we have a confused record of marchings and counter marchings, in which Cuttack was the prize for which the contending parties struggled. Not long after the departure of Alī Vardi Khān, Raghujī Bhonsla suddenly burst upon Orissa and appeared under the walls of the fort, where the garrison sustained a vigorous siege for about a month. The citadel was however ill furnished for a long defence, provisions ran short, and at last the commandant capitulated and the Marāthās took possession of the city. In 1746 Raghujī Bhonsla, who had in the meantime been busy with his raids in Bengal, retired to Berār, and next year Alī Vardi Khān determined to conduct a vigorous campaign against the Marāthās in Orissa and to recover the capital. Reinforcements were sent from Berār by the Marāthās, but Alī Vardi Khān, making a forced march, compelled them to surrender the fort after a siege of 15 days. It soon passed again into the hands of the Marāthās on the cession of Orissa to them in 1751, and they held undisputed possession of it till the advent of the British in 1803. The Marāthās had shut themselves up in the fort, and the small invading force entered Cuttack without meeting any opposition on the 8th October 1803. They at once started to erect batteries and make the approaches. The fort, strongly built of stone and surrounded by a wet ditch, varying from 35 to 135 feet in breadth, had only one entrance, with a very narrow bridge leading over the ditch to it. The batteries were completed by the night of the 13th October, five hundred yards from the south face of the fort, and they commenced firing early the following morning. By 11 A.M. all the defences had been knocked to pieces, and the guns of the fort silenced. The storming party, consisting of a detachment from His Majesty's 22nd Regiment and the Madras European Regiment, 400 sepoys from the 20th Bengal Native Infantry, the 9th and 19th Madras Native Infantry, and some artillery, with a six-pounder to blow open the gate, advanced to the attack. The bridge was quickly passed, under a heavy fire from the fort, but it was nearly forty minutes before the wicket was blown sufficiently open to admit one man. The Europeans passed in singly, but with such rapidity, that, notwithstanding the resistance at the inner gates, they entered with the garrison, who after a very severe loss abandoned the fort.

Bārabāti fort. The fort of Bārabāti was built by Mukunda Deva, the last Hindu king of Orissa (1560—68), and was apparently a castle of grey granite with nine lofty courts. In the Ain-i-Akbarī it is described as a fine palace consisting of nine courts, the first of which was used for the elephants, camels and horses; the second was a store-house for the artillery and military stores, and also contained quarters for the guards and other attendants; the third was occupied by porters and other watchmen, the fourth by artificers, and the fifth by the kitchens. The sixth contained the Rājā's public apartments, the seventh was used for the transaction of private business, the eighth was the zanāna, and the ninth contained the Rājā's own sleeping apartments. It was here that the Mughal Sūbahdārs held their court, and fortunately we have a description of its splendour in the account of William Bruton, who visited it with Ralph Cartwright in 1633. He was much impressed with the magnificence and pomp of the stately Court of Malcandy, as he calls it, Malcandy being apparently a corrupt form for Mukunda Deva. "The English travellers," writes Mr. Wilson in 'The Early Annals of the English in Bengal,' "reached the place from the east over a long narrow causeway, and were conducted through a labyrinth of buildings to the court of public audience. Here Bruton and his companions awaited the coming of his Highness, and found themselves objects of much curiosity. At last the word came that the nabob was approaching. The place was forthwith spread with rich carpets, gold pillars being placed at the corners to hold them down, and in the middle a red velvet bolster for his Highness to recline against. Then, preceded by his brother, a comely man carrying a sword, accompanied by fifty grave-looking courtiers, and greeted on all sides with low prostrations, came the Mogul Governor, a fair and stately personage, leaning his arms upon two of his attendants. This was Aghā Muhammad Zamān, a Persian grandee, born in Tahrān, who was in high favour with the Emperor Shāh Jahān, and had recently been sent to Orissa to wage war against the King of Golkonda. He very affably inclined his head towards Mr. Cartwright, who was presented to him by Mirzā Momin, and, slipping off his sandal, offered 'his foot to our merchant to kiss, which he twice refused to do, but at last he was fain to do it.' Then the nabob and the whole court sat down cross-legged. The English merchant brought forth his presents, and made his requests to the nabob for trading privileges. But by the time he had reached the end of his story, the King's almoner gave the signal for prayers, and the whole company knelt down with their faces towards the setting sun. Prayers being ended, and business laid aside, the palace was soon ablaze with countless wax tapers which the attendants lighted up with great ceremony."

Even as late as the beginning of the 19th century, the citadel must have been an imposing sight, to judge from Mr. Stirling's description of it. "The only monument," he writes, "of the Gajapati Rājās which their ancient capital exhibits is the fortress of Bārabāti, built probably in the 14th century by Rājā Ananga Bhīm Deo. Some ascribe its erection to Telinga Mukund Deo, the last of the independent sovereigns of Orissa, and others refer it back to a period as early as the times of the Kesari dynasty. However that point may stand, its square sloping towers or bastions, and general style, bespeak clearly a Hindu origin. The Muhammadan or Marāthā governors added a round bastion at the N.-W. angle, and constructed the great arched gateway in the eastern face, which alterations are alluded to in a Persian inscription, giving for the date of the repairs and additions, the fourth year of the reign of Ahmed Shāh or A.D. 1750. The fort has double walls built of stone, the inner of which enclose a rectangular area measuring 2,150 by 1,800 feet. The entrance lies through a grand gateway on the east, flanked by two lofty square towers, having the sides inclining inwards, from the base to the summit. A noble ditch faced with masonry surrounds the whole, measuring in the broadest part two hundred and twenty feet across. From the centre of the fort rises a huge square bastion or cavalier supporting a flag-staff. This feature, combined with the loftiness of the battlements on the river face, give to the edifice an imposing, castellated appearance, so much so that the whole when seen from the opposite bank of the Mahānadī presented to the imagination of Mr. La Motte, who travelled through the province in 1767 A.D., some resemblance to the west side of Windsor Castle. No traces of the famous palace of Rājā Mukund Deo nine stories in height, mentioned in the Ain-i-Akbarī, are to be found within the walls of fort Bārabāti, but the fragments of sculptured cornices, etc., whichhave been dug up at different times, and more especially a massive candelabra, or pillar furnished with branches for holding lights, formed of the fine grey indurated chlorite or pot stone, are probably the remains of some large and splendid edifice."

There is little in the present appearance of the fort, or *kilā*, as it is generally called, which answers to the above description. The Public Works Department, in early vandal days, stripped the old buildings for the sake of their stone, which they used for the False Point light-house and other buildings as well as for metalling the roads, and thus converted the fort into an unsightly series of mounds, and the ground within the moat into a wilderness of stone pits. The old *kilā* now contains the buildings of the Station Club, and the hospital, magazine and other buildings of the wing of the Madras regiment which is stationed at

Cuttack. The great arched gateway to the east and an old mosque, named after Fathi Khān Raham, are the only objects of antiquarian interest which remain intact.

Lālbāgh. In spite of all its attractions, the Mughal and Marāthā Governors did not reside in the fort, but in a palace at Lālbāgh on the bank of the Kātjurī. According to William Bruton, "although the palace of the nabob be so large in extent and so magnificent in structure, yet he himself will not lodge in it, but every night he lodgeth in tents, with his most trusty servants and guards about him; for it is an abomination to the Moguls (which are white men) to rest or sleep under the roof of a house that another man hath built for his own honour. And therefore he was building a palace, which he purposed should be a fabric of a rest and future remembrance of his renown: he likewise keepeth three hundred women, who are all of them the daughters of the best and ablest subjects that he hath." The Commissioner's residence now occupies the site of this palace.

The only other building of antiquarian interest in the town is the Kadam Rasūl, which Stirling describes as an antique-looking edifice standing in the midst of a fine garden, which contains certain relics of the Prophet commissioned from Mecca by the Nawāb Nāzim Shujā-ud-dīn Khān, or his son Muhammad Taki Khān, the latter of whom lies buried within the enclosure. From an inscription *in situ* it appears that it was built by the Nawāb Shujā-ud-dīn Muhammad Khān in the reign of Shāh Alam (1707—12). It is an ordinary brick building, covered with whitewash, of no special merit, inside which the foot-prints of the Prophet are kept in a basin of water. This holy water is given to persons visiting the shrine and is used for curing diseases. The building is pleasantly situated in a large garden, where there are many tombs of former custodians of the shrine; it is supported by a large endowment and is well looked after.

The cemetery contains tombs with inscriptions dating back to 1808, the most interesting of which is one, dated 1811, with an epitaph to the memory of one Turner, an artificer, which runs:—

"My hammer and anvil lie declined,
My bellows too have lost their wind,
My time is spent, my glass is run,
My last nail's drove, my work is done."

Deuli.—A village in the Jājpur sub-division, situated 2 miles west of the police-station of Dharmshāla. It contains a small temple sacred to Gokarneswar, which is picturesquely situated on the bend of the river Brāhmanī round the Deuli hill. The roof of the pillared hall has fallen, and the temple is in a bad state of repair. In front of it grows a banyan-tree, at the foot of which is a life-sized monolithic image of the four-handed Vishnu, which was recovered some years ago from the river-bed.

False Point.—Cape, harbour and light-house in the Kendrāpāra sub-division, situated on the north of the Mahānadī estuary, in latitude 20° 20′ N., and longitude 86° 47′ E. It derives its name from the circumstance that it was often mistaken by ships for Point Palmyras one degree further north. The harbour consists of an anchorage, land-locked by islands and sand-banks, with two navigable channels inland. The light-house stands on the point, which screens it from the southern monsoon, in latitude 20° 19′ 50″ N, and longitude 86° 44′ 30″ E. The anchorage is protected by two sandy reefs, called Long Island and Dowdeswell Island, and is completely land-locked by the latter. Point Reddie on the Dowdeswell Island shelters the entrance; and further in lies Plowden Island, for the most part a low jungly swamp, with a limited area of high ground suitable for building purposes and possessing good drinking-water. The harbour is safe and roomy, the channel properly buoyed, and a soft mud bottom prevents injury to vessels running aground. The port is open throughout the year, but during the last few years the inner harbour has silted up, and vessels drawing over 14 feet of water now have to lie at the outer anchorage. A Port Officer and an Assistant Superintendent of Customs are stationed here.

Two separate channels lead inland from the anchorage, the Jambu river on the north, and on the south the Bākud, a short deep branch of the Mahānadī. Bars of sand intervene between the anchorage and these channels, but at full tide cargo boats and steamers enter with ease. Several tidal creeks, navigable by country boats throughout the year, also connect the harbour with the Dhāmra and Brāhmanī rivers on the north and with the Devī on the south.

The Jambu channel is a winding stream dangerous to navigation when freshets come down. A bar stretches across its mouth for about three quarters of a mile, with one foot of water at the lowest tide; after this, the channel gradually deepens to 10 feet at the lowest tide, and still higher up to 18 feet. Towards Deulpāra some 12 or 15 miles from its mouth, the Jambu shoals and narrows to such an extent that navigation becomes dangerous for heavily-laden country boats. The entire course of this channel is through a desolate country, which during floods forms one large sea or jungle-covered swamp. The Bākud creek is the more direct of the two channels for navigation. A bar about 1,000 yards long stretches across its mouth, which is dry during the last quarter of the ebb, but at full tide cargo boats and steamers enter freely. Beyond the bar a channel of 2 feet is obtained, gradually deepening to 8, then shoaling again to 2, and eventually deepening into an excellent channel of 19 to 20 feet up to its junction with the Mahānadī, a distance of 16 miles. It was on this creek that Government established its rice depôt during the great famine of 1866.

The trade of False Point is chiefly with other Indian ports, but a considerable export trade in rice and oil-seeds is also carried on with Colombo and Mauritius. The British India Steam Navigation Company make the place a regular port of call and practically have the monopoly of the trade. Formerly, it was a busy port, and in 1877-78, it was entered by 202 vessels with a tonnage of 141,000 tons, the value of its trade being over 72 lakhs. After that year it steadily declined, and its trade still further diminished owing to the opening of the East Coast Railway in 1899, when it was visited by 28 vessels, the value of its trade being only 11 lakhs. The effects of the competition of the railway appear however to have been only temporary, and the port is now steadily gaining its former position; in 1904-05 altogether 37 vessels with a tonnage of 80,000 tons entered the port, the value of the imports being Rs. 3,358 and of the exports Rs. 22,71,606.

False Point was formerly considered very unhealthy, but the malaria to which it owed this evil reputation has to a great extent disappeared. Though the inner anchorage has silted up within the last few years, the outer anchorage, which contains a depth of 24 feet of water, provides protection against the south-west monsoon; and despite the opening of the East Coast Railway, the volume of its trade has increased of late years, thus proving its value as a port, and to some extent fulfilling the anticipations of its utility.

Hariharpur.—(Literally the city of the tawny one and the grasping one, *i.e.*, the city of Vishnu and Siva.) A village adjoining Jagatsinghpur, on the Alankā, about 25 miles from Cuttack. Till the beginning of the 19th century the two villages were called Hariharpur; the place now goes by the name of Jagatsinghpur, owing to the greater importance of the latter village. Hariharpur is of great historical interest as being the site of the first factory established by the English in Bengal. Ralph Cartwright and his two companions, Colley the second merchant, and Bruton the ship's quartermaster, stopped here in 1633 when on their way to Cuttack to obtain a permit allowing them to trade in Orissa. Here they met with as good a welcome as at Bālikudā, 11 miles march away, the Governor of which had helped them on their way with horses to ride and coolies to carry their baggage, and had escorted them with "music played most delicately out of tune, time and measure." According to Bruton, a nobleman named Mersymomeine (Mirzā Momin), "one of the king's greatest noblemen and his most dear and chiefest favourite," met them "at a great *pagoda* or *pagod*, which is a famous and sumptuous service and worship there used; and giving them a warm welcome, entertained them with a very great feast or costly collation." This great pagoda Bruton calls a stately and magnificent building, but what it actually was is not certain. There is an old temple of Siva at Hariharpur, known locally as Somnāth, but from the fact that Mirzā Momin and his followers stopped in the pagoda, it has been suggested that it may have been a pavilion erected for royal encampments. After obtaining from the Mughal Governor the concession they had demanded, the English returned to Hariharpur on the 10th May 1633, and "hosted" in the house of their interpreter. They at once started to found a factory, and Bruton's quaint description of their proceedings shows what keen men of business they were. On the 11th May, the day after they arrived, he says, "we went to the Governor of the town and showed him our *fermand*, or commission from the king: the governor made a great *salame*, or court'sy, in reverence unto it, and promised his best assistance and help in anything that he could do; and there the said governor had a small present given to him. The fourteenth day, the two merchants went abroad, and found out a plot of land fitting to build upon; then they laid the king's *deroy* on it and seized upon it for the Company's use; and there was no man that did or durst gainsay them for doing the same. The fifteenth day they hired workmen and labourers to measure the ground and to square out the foundation of the house, and likewise for the wall, which was one hundred conets square, which is 50 yards, every conet being half a yard or a foot and a half; and it behoved us to make haste for the time of the great rains was at hand. The sixteenth day they laid the foundation of the walls, being 9 feet thick: much haste was made and many workmen about it; but this our first work was but labour lost and cast away, for it came to nothing. For on the eighteenth day the rains began with such force and violence that it beat down all our work to the ground and washed it away as if there had not been anything done: this storm continued without ceasing (day and night), more or less, three weeks complete."

The building had to be begun again, and when it was finally completed, the English proceeded to carry on a trade in the silks for which Hariharpur was then noted. The factory however soon fell in decay, as the river silted up and cut off access from the sea; in 1641 it was on the point of dissolution, and soon afterwards it was abandoned.

Harispurgarh.—A village situated at the extreme south-east of the district, at the mouth of the Patuā. It was here that the English first landed in 1633 when they came north from Madras in order to exploit Bengal. Whether by accident or not, the Portuguese appear to have got wind of their design, and set upon them as they lay at anchor. The following account of their short stay here is taken from Bruton's diary of his voyage (reproduced in Wilson's Early Annals of the English in Bengal, Vol. I):—"The twenty-first of April, being then Easter-day, we were at anchor in a bay before a town called Harssapoore; it is a place of good strength with whom our merchants hold commerce with correspondency. This twenty-first day in the morning Mr. Ralph Cartwright sent the money ashore to the Governor of Harssapoore to take it into his safe-keeping and protection until such time he came ashore himself. So presently there came a Portugal frigate fiercely in hostility towards us, but we made ready for their entertainment and fitted ourselves and the vessel for our best defences; but at last they steered off from us, and, upon our command, she came to an anchor somewhere near us, and the master of her came on board of us, who being examined whence he came and whither he was bound, to which demands he answered nothing worthy of belief as the sequel showed: for he seemed a friendly trader, but was indeed a false invader (where opportunity and power might help and prevail); for, on the 22nd day, Mr. Cartwright went ashore to the Governor of Harssapoore; and on the 24th day, the said master of the frigate (with the assistance of some of the ribble-rabble rascals of the town) did set upon Mr. Cartwright and Mr. Colley, where our men (being oppressed by multitudes) had like to have been all slain or spoiled, but that (*Lucklip*) the *rogger* (or vice-king there) rescued them with two hundred men.

"In this fray Mr. Thomas Colley was sore hurt in one of his hands, and one of our men much wounded in the leg and head; their *nockada*, or India pilot, was stabbed in the groin twice, and much mischief was done and more intended; but by God's help all was pacified. The twenty-seventh day of April we took leave of the governor and town of Harssapoore (I mean three of us); namely, Mr Cartwright, William Bruton, and John Dobson, leaving Mr. Colley and the four men with him, till news could be sent back to them from the nabob's court at Cutteke or Malcander, of our success and proceedings there with our other goods; for he is no wise merchant, that ventures too much in one bottom, or that is too credulous to trust Mahometans or Infidels."

Jājpur sub-division.—North-western sub-division of the district, lying between 20° 39′ and 21° 10′ N., and 85° 42′ and 86° 37′ E., and extending over 1,115 square miles. Its population was 560,402 in 1901 against 525,910 in 1891. The west of the sub-division lies on the fringe of the Chotā Nāgpur plateau, and this portion is very sparsely populated. Towards the east, which consists of a fertile highly cultivated plain, the density increases, the figure for the whole sub-division being 503 persons to the square mile. It contains one town, Jājpur, its head-quarters, and 1,580 villages. In 1903-04 the area under cultivation was returned at 586 square miles, of which 50 were irrigated from the canals, and the area of culturable waste at 43 square miles.

Jājpur town.—Head-quarters town of the sub-division of the same name, picturesquely situated on the right bank of the Baitaranī river in 20° 51′ N. and 86° 20′ E. Population (1901) 12,111. It is also the head-quarters of a Public Works Department sub-division, and, besides the usual public offices, contains a sub-jail, an English High school, and a charitable dispensary with 4 beds for male and 2 beds for female patients. The town is 14 miles from the Jājpur Road station on the Bengal-Nāgpur Railway, but as there is no good road, the journey cannot be conveniently performed except in a *pālki* or on horse-back, but neither horses nor *pālkis* are available at the station without previous arrangement. A new road from the Vyās Sarobar station has recently been constructed, and the town is also connected with the Grand Trunk or Jagannāth Road by an unmetalled road, 9 miles long.

Legendary history. The name Jājpur (Jajnapura, the town of the sacrifice) is connected with the legend that Brahmā brought 10,000 Brāhmans from Kanauj for the performance of a ten-horse sacrifice (*Dasāsvameda Jajna*). Among the gods who thronged to this august sacrifice came Holy Mother Ganges (Gangā); and tradition asserts that ever since those solemn rites she has sent an offshoot of her waters through the bowels of the earth into Orissa, which emerges as the sacred Baitaranī river, the Styx of the Hindus. Leaving aside the mythical Brahmā, it appears possible that a great ceremony was performed with the object of reviving the Brahmanical faith and of supplanting Buddhism, which had obtained a firm hold on the country. Pure Brāhmans were evidently, therefore, imported from Kanauj, the greatest stronghold of the Brahmanical faith in Northern India. The king with whom the revival of Brahmanism in Orissa is usually associated had his capital at Jājpur, and the great ceremony, which the inventive genius of later mythologists attributes to Brahmā, may have been his work. A somewhat similar story is current in Bengal, where five Brāhmans, the ancestors of the modern Kulins, are said to have been brought from Kanauj by king Adisūra. There are traditions that the Brāhmans who congregated at Jājpur for the great sacrifice and their descendants gradually spread over the rest of Orissa, and it is noticeable that the town and its neighbourhood are still inhabited by large colonies of Brāhmans, holding royal grants called *Sāsans.*

Another tradition connects Jājpur with the Gayā legend, according to which Brahmā induced Gayā Asura (a respectable pagan monster of great sanctity, whose only fault was that he would save sinners from perdition) to lie down for a feast to be held on his body ; and having done so, placed a large stone on him to keep him there. Gayā, however, struggled so violently that it was necessary, when force failed, to persuade him to be quiet, which was done by a promise being made that the gods would take up their abode on him permanently, and that any one who made a pilgrimage to the spot, and performed certain ceremonies, should save himself and his ancestors from the penalties of the Hindu place of torment. Such was the vast bulk of the monster that when stretched on the ground his head rested at Gayā, and his navel at Jājpur ; and a sacred well, a few feet deep, called the Gayā Nābhi (navel) still commemorates his fall. Here the pilgrims make offerings of *pindas* or rice-cakes as an expiation of the sins of their ancestors, in the same way as in the great pilgrim city of Gayā. Dr. Rājendralāla Mitra sees in this legend an allegory of the triumph of Hinduism over Buddhism, and suggests that the area covered by the body of Gayā perhaps alludes to the territory in which Buddhism prevailed; and it may be that this legend, like that of the great sacrifice from which the town obtains its name, points to the former prevalence of Buddhism and to its disappearance before the growing popularity of the Brahmanical faith.

The Gayā legend is a Vishnuite legend, in which Vishnu plays an important part; and it is interesting to notice that side by side with it is a well-known Sivaite legend which explains the sanctity of Jājpur as a sacred city of the bloody goddess, Kālī. According to this myth, Siva became so disconsolate after the death of his wife Satī that he wandered for ages through the world carrying her corpse. To put an end to his despondency, Vishnu cut up the corpse with his celebrated *chakra* into 51 fragments, which falling in as many places made the 51 places of pilgrimages devoted to the goddess of destruction. A temple at Jājpur, containing the image of Satī under the name of Birajā or the passionless one, now marks the sacred spot on which one of these fragments fell. The present building is comparatively recent and cannot be of an earlier date than the 14th century; but the site is very old, and from the mention of "Birajā Kshetra" in the Mahābhārata it has been inferred that it was a sacred spot as early as the 2nd or 3rd century A.D.

Jājpur is said to have shared with Bhubaneswar, the honour of being the capital of Orissa till the 10th century A.D., when the seat of Government was removed to Cuttack, the present capital of the Province; and it has been suggested that the name is derived from Jajātipura, as Jajāti Kesarī, the first Kesarī king of Orissa, it is said, held his court here in the 5th century and built himself a castle and palace in the town. It did not, however, lose all its importance, as it continued to be one of the five *Katakas* or fortified capitals of the kings of Orissa, and at a later period, Muhammad Taki Khān, the Deputy of the Nawāb Shujā-ud-dīn, held his court and built a palace here on the site now occupied by the sub-divisional buildings. His palace was pulled down by one of the Marāthā *āmildārs*, who used the stone to build his own mansion and the temple of Gobindjī at Bhogmādhab, a mile from the town.

Tradition says that the last great battle between Mukunda Deva and the Afghān conquerors was fought at a place, called Gobira Tikri, about 4 miles to the north-east. The place is still dreaded, as it is believed that whole armies are lying sunk in the adjoining marshes, where they still beat their drums and blow their trumpets at dead of night.

Monuments.

Jājpur contains within its limits relics of almost all the phases through which image worship has passed in Orissa. Leaving the temple of Birajā, one finds shrines of Siva scattered all over the town and its vicinity, of which the most important are those of Akhandaleswar, Agneswar, and Trilochaneswar. The first two at least must be of some antiquity, as they are mentioned in the Mādalā Pānji, or palm-leaf chronicles of the temple of Jagannāth, as having received grants from King Ananga Bhīma Deva. The temple of Akhandaleswar contains among others a well-carved image of a small naked figure with a placid countenance, which is evidently a Jaina Tīrthankara; and the lingam of Agneswar is believed to change its colour every quarter of the day.

On the bank of the sacred Baitaranī, stand, side by side, a modern temple of Jagannāth erected in the time of Raghujī Bhonsla, containing an image removed from the fort at Solampur, and a much older temple of Kālī of the usual Orissaic type. To the east of Kālī's temple overlooking the bed of the sacred river, is a raised gallery containing eight life-size monolithic statues of the seven Mothers, Indrānī, Vārāhī, Vaishnavī, Kumārī, Yama Mātri, Kālī, and Rudrānī, and of the Nrīsingha incarnation of Vishnu. Stirling writes of them as follows:—"They are said to have been recovered lately out of the sand of the river, where they were tossed by the Mughals on their shrines being destroyed, by a *mahājan* of Cuttack who built the edifice in which they are now deposited. The figure of Kālī is sculptured in a very spirited manner; she is represented with an axe in one hand and a cup full of blood in the other, dancing in an infuriated attitude after the destruction of the giant Rakta Vija, and trampling unconsciously on her husband Mahādeva, who, as the fable runs, has thrown himself at her feet to solicit her to desist from those violent movements which were shaking the whole world. That of Yama Mātri, the mother of Yama, is also a very striking and remarkable piece of sculpture. Her form is that of a hideous decrepit old woman seated on a pedestal, quite naked, with a countenance alike expressive of extreme age and that sourness of disposition which has rendered her proverbial as a scold."

On an island in the middle of the river stands the temple of Varāhanāth, the boar incarnation of Vishnu, which is said to have been repaired by King Pratāparudradeva (1496—1530 A.D.) and to have been visited by Chaitanya, the Bengal apostle of Vaishnavism, about A.D. 1510. The temple is approached by a flight of steps, the name of which, Dasāsvameda Ghāt, commemorates the great sacrifice mentioned above. Within the compound of the Subdivisional Officer's quarters are four colossal images named Vārāhī, Chāmundā, Indrānī, and Kalijuga. They once adorned the colonnade of the *Mukti mandapa* or conclave of Pandits, but were flung down by the conquerors, who broke up the remainder, and made them into cannon-balls. The last is a mutilated representation of the Iron Age, which was brought here from Sāntmādhab, about a mile off, where it was lying buried, and the first three are members of the Hindu group known as the seven mothers, which were removed from the adjoining *āsthān* of a Muhammadan *pir*, to which the bigoted fury of the Muhammadans had consigned them.

Adjoining the compound of the sub-divisional office stands the mosque of Abu Nasīr Khān, erected towards the end of the 17th century with materials obtained from the demolished Hindu temples. It has four eminarts on the east face, and three domes on the roof, which are mere coverings to the three flat ground domes forming the roof of the interior. There are openings to the interior of the domes on the roof, and very possibly they were used, and are used, as places of concealment. There is a Persian inscription over the centre doorway, of which the following is a translation:—"In tho timo of Aurangzeb, whose splendour reaches the stars and will remain as long as the stars endure, in the time of the Nawāb whose virtues are altogether beyond praise or description, the Nawāb established in the city of Jājpur a mosque of such magnificence that the domes of it make the sky conceal itself. If you desire to hear the messages of the angels, spend a night in it. Abu Nasīr Khān reigned when the mosque was erected:—then was the time of Abu Nasīr Khān."

A few yards off from the road leading from the bazar to the temple of Birajā, a massive stone pillar, known as the Chandeswar pillar, exquisitely chiselled and well proportioned, marks the site of a temple of proportionate dimensions, every trace of which has been obliterated. The pillar is 22 feet high, standing on a pedestal of three enormous blocks of stone, each about 5 feet long, 5 feet broad, and 2 feet thick. The monolith itself is 3½ feet square at the bottom, bevelled off for some inches at each corner, and fluted above for a height of about 20 feet. On the top of the monolith is another block of stone, wrought into a regular capital, on the lower portion of which garlands are sculptured. The stone is then cut into the shape of a lotus calyx which supports the upper portion, a square of about 4½ feet. On this was the figure of Vishnu's vulture, which was pulled down by the great Musalmān iconoclast Kālā Pāhār, and now rests in a small temple about half a mile off. The fury of the iconoclast was however wasted in the attempt to pull down the column itself by means of chains and teams of elephants. Holes were drilled for the chains, and the column was moved an inch or two from its position on the pedestal, but it still rears its lofty head in defiance of the elements to which it is exposed.

The visitor cannot fail to notice more patent marks of Musalmān fury in the disfigured faces and broken arms of images and the broken capitals and pillars found on all sides To quote the graphic account of Sir William Hunter:—"Whatever Musalmān bigotry could destroy has perished; and the grave of an Afghān iconoclast, quarried out of Hindu shrines, now forms the most conspicuous monument in the metropolis of the Sivaite priests. The Muhammadans of the sixteenth and seventeenth centuries stabled their horses in the Hindu palaces, and tore down the great temples, stone by stone, to build royal residences for their own chiefs. At first the Orissa deities, who became the demons of the Musalmāns, as the gods of Greece and Rome furnished devils to primitive Christendom, resisted by signs and portents. But there came a saint in the Afghān army, named Alī Bukhar, a follower of Kālā Pāhār, whose detestation of the infidel had transported him from Central Asia to the Bay of Bengal, and whose piety (or persecution) cowed the evil spirits of the bygone creed into silence. He threw down the colossal statues of the Hindu gods, and for nearly three centuries they have lain prostrate under his mystical spells. The great high place of Sivaism resounded with the Friday prayers and the daily readings of the Korān; and a curious document, dated upwards of two hundred years ago, still enjoins the Jājpur authorities to pay the cost for lamps to the Musalmān family in charge of the public ministration of Islām." This Alī Bukhar, legend relates, had his head cut off in the final assault in Fort Bārabāti at Cuttack, but his headless trunk spurred his horse till it reached Jājpur. Here the body was buried on the high terrace where his tomb still stands, his horse being buried in a separate grave beside him. It is characteristic of this iconoclast that his tomb should be built on the site of the *Mukti mandapa*, which was destroyed by the conquerors.

The Tentulimal bridge.

Not far from Jājpur is a peculiar bridge of 11 arches, called the Tentulimal bridge, which appears to belong to an age ignorant of the use of the arch, and to be one of the bridges erected by the early sovereigns of Orissa. The arches are formed entirely by corbelling, *i.e.*, by making each successive layer of masonry overlap the layer below, until the two piers come to within a foot of each other at the top. On this space a long narrow block of stone is laid as a sort of keystone, over which enormous blocks of stone, some of them 4½ feet long and half that in breadth, are laid transversely, apparently with the object of making the upper layer of the piers on each side grip the keystone. The whole bridge is about 240 feet long and 32 feet broad, and each of the piers is about 10 or 12 feet wide; the abutments at each end are of laterite, but in the centre the masonry is of coarse red granite. There are two points, however, in the bridge which tell somewhat against its antiquity. The first is that in various parts of the piers and under the opening arches one notices stones which have evidently formed part of another and older structure. Here and there are bits of chequered carving, and also a number of stones with bas-reliefs on them, disposed with a certain regularity in the centre of each pier, which evidently have come from an older structure and have been fitted into this. One of these is a carving, probably intended for a representation of Buddha in a sitting posture, which has the square broad face, long ears, and heavy head and feet one often sees in figures of Buddha. The other point is the two kinds of stone used. The laterite of the abutments and of one or two of the smaller side-arches, and the coarse granite of the centre arches, would indicate that the masonry of the centre arches is of more recent construction than that at the sides. In any case, however, the bridge must have been constructed before the Mughal conquest of Orissa, and possibly the explanation of the difference in the stone is that, the centre arches of laterite having fallen in, the ruins of some Buddhist temples were used to rebuild it. An interesting account of the antiquities of Jājpur and of the traditions current there will be found in an Account of the Antiquities of Jājpur in Orissa, by Babu Chandra Sekhar Banerji, J.A.S.B., Vol. XL, No. 2, 1871.

Kanikā.—Kilā Kanikā, with an area of 440 square miles, is the largest estate on the Cuttack revenue-roll, but out of its total area, 175 square miles are situated within the geographical limits of the district of Balasore. The formation of the tract is deltaic. The lower portion close to the sea-coast consists of low dense, marshy jungles, which become thinner and higher as they recede from the sea. Higher up are arable plains, the lower portions of which are subject to salt-water floods during storms and cyclones, and the upper to inundation by the many branches of the Baitaranī and Brāhmanī rivers. The crops are always liable to be destroyed, and it must be reckoned as one of the parts of the country most liable to famine.

It is said that this part of the country was peopled by aboriginal tribes, ruled over by several petty Rājās until about the year 1200 A.D. when a brother of the ruling chief of Mayūrbhanj established himself in possession of the portion of the *kilā* now known as Ilākā Chamuka. By conquest and marriage his successors added to their property, and at the time of the British conquest the Rājā, Balabhadra Bhanja, held not only the four *ilākās* of the present estate of Kanikā, but the zamīndāri of the large estate of Utikan. The first Commissioners in 1803 confirmed his *peshkash* of 84,840 *kāhāns* of cowries, equivalent to Rs. 20,408.

In 1805, owing to the Rājā's malpractices, he was imprisoned and the estate was held under *khās* management. Next year he was reinstated, and on his death, in 1813, he was succeeded by his two sons. In 1845, during the minority of Balabhadra Bhanja's great-grandson, the estate came for four years under the Court of Wards, by whom a settlement was made which raised the *jamā* to over Rs. 82,000. The estate was again brought under the management of Government in 1862, on the application of the late Rājā Padmanābh Bhanja; and as he was declared insane in 1865, it continued to be held by the Court of Wards until his death in 1891, and was then managed by the Court on behalf of his adopted son, Nripendra Nāth Bhanja, and after his death in 1895 on behalf of the minor (adopted by the Rānī) Rājendra Nāth Bhanja, who is a son of the Rājā of Aul. It was released in 1902, when the ward attained his majority.

Kendrāpāra sub-division.—North-eastern sub-division of the district, Bengal, lying between 20° 18′ and 20° 48′ N., and 86° 15′ and 87° 1′ E., and extending over 977 square miles. Its population was 467,081 in 1901 against 429,770 in 1891. It is bounded on the east by the Bay of Bengal, and the tract along the coast is very sparsely populated. The density rises towards the west, and the figure for the whole sub-division is 478 persons to the square mile. It contains one town Kendrāpāra, its head-quarters, and 1,338 villages. In 1903-04 the area under cultivation was returned at 512 square miles, of which 47 square miles were irrigated from the canals, and the area of culturable waste at 137 square miles.

Kendrāpāra town.—Head-quarters town of the sub-division of the same name, situated in 20° 30′ N. and 86° 25′ E. Its position on the Kendrāpāra canal in the heart of a rich grain-producing country gives it a considerable trade, and it is connected by road with Cuttack, Jājpur and Chāndbāli. Besides the usual public buildings, Kendrāpāra possesses a good school and dispensary, a sub-jail, and a public library which has lately been opened for the circulation of English and vernacular literature. The town is divided into three portions, one between the canal and the river Gobri, a second portion north of the river, and a third south and east of the canal. The area within municipal limits is 2½ square miles, and the population, according to the census of 1901, is 15,245.

Kujang.—Kilā Kujang is one of the great *kilās* of the district, and has a total area of 370 square miles. It consists of two distinct tracts, the first a marshy and almost uninhabited strip along the sea-coast, and behind this low-lying arable lands intersected by innumerable streams and tidal creeks, which both inundate the land and supply means of irrigation in the cold weather. As in Kanikā, the harvest is liable to be destroyed by storms and cyclones, as well as by floods in the Mahānadī river.

The original Rājās of Kujang were descended from Mallik Sendh, Rājā of Dhobaigarh, who lived in the 17th century and, with the help of the Rājā of Kanikā, fought with and subdued the chiefs of the neighbouring *garhs*, amalgamating them all under the name of Kujang. At the time of the British conquest, Gangādhar Sendh was in possession and executed an *ekrārnāmā* for the payment of a *peshkash* of 14,011 *kāhāns* of cowries. In 1812 his son tried to instigate the Rājās of Kanikā and Khurdā to join him in conspiracy against the British authority, but the attempt being detected, he was dispossessed and imprisoned; and his brother Bīrabhadra Sendh ruled in his stead and executed a new agreement for the payment of Rs. 7,501. The estate was seriously affected by the famine of 1866, and in 1869 it was sold for debt and bought by the Mahārājā of Burdwān. On the death of the Mahārājā, the estate came under the Court of Wards, and between 1887 and 1892 a cadastral survey and settlement was carried out under Act X of 1859.

Mahāvināyaka Hill.—A peak of the Bārunībunta Hills in the Jājpur sub-division, situated in 20° 42′ N. and 86° 6′ E. The hill is covered with jungle, and is seldom visited by any but pilgrims. It was probably from the beginning a Sivaite place of worship, no signs of Buddhism being traceable. On the northern slope of the hill, about 400 feet above the level of the surrounding country, there is a monastery, occupied by Vaishnavas, who have evidently superseded the original Sivaits; and close by is a modern temple built on a base of cut stone, which is all that remains of the old shrine which once stood here; the walls and pyramids were destroyed by the Muhammadans and appear to have been rebuilt subsequently. The principal curiosity of the place is a massive piece of rock, known as the god Mahāvināyaka, over which the modern temple has been built. The rock is over twelve feet in circumference, oval at the top, and has three faces in front. The middle one bears a tolerable resemblance to the head and trunk of an elephant, and is accordingly worshipped as Ganesh or Vināyaka; the right face of the rock is revered as a representation of Siva; and, according to popular belief, a knot over the left face represents the bound-up tresses of the goddess Gaurī. The rock is accordingly worshipped as the union of the gods Siva and Ganesh and of the goddess Gaurī. About 30 feet higher up there is a waterfall, which supplies water to the temple and pilgrims, and a few steps above this fall are some images of Siva, called the *Ashta Lingam* from their number. On the south side of the hill are the ruins of a fort known as Teligarh; the walls and inner rooms are of laterite and the doorways of gneiss.

Naltigiri Hill.—One of the hills of the Assia range in the Jājpur sub-division, but separated from it by the Birūpā river, situated in 20° 35′ N. and 86° 15′ E. The hill consists of two spurs, the smaller one called Arasuni and the larger, which extends from east to west for about half a mile, Nanda. On the Arasuni spur is a monastery known as Abhyāgatāsrama, and in its neighbourhood is a group of small modern temples, one of which goes by the name of *Gumphā* or cave. An old flight of steps now in ruins leads to the other spur Nanda, which consists of two peaks of unequal height with a small pass between. On this pass is a small flat-roofed temple of the *guru* Vāsuli Thākurānī, which is of modern date but was evidently built of old materials on the foundation of an older structure. The building consists of a porch and a cell surmounted by a small pyramidal tower; the roof of the porch has given way, but that of the cell still stands. It has no columns, and is formed of solid walls, with niches in the interior which contained 5 images of Bodhisatwa, or Ananta Purushottama, as people on the spot call them. One of these images still stands, but the other four are lying on the ground with their faces upwards. The figures are about 5 feet high, each holding in the left hand a lotus with a long stem, cut in high relief. The one standing and two of those lying on the ground bear inscriptions in the Kuthila character. Near the higher of the two peaks is a coarse uncouth figure of Padmapāṇī, known to the villagers as Ghantiasuni, and on the top is a terraced stone platform, which, according to tradition, is the *baithak* or seat of Rājā Vasukalpa Kesarī, but which there can be little doubt represents the remains of a Buddhist stūpa. Ascending the other part of the hill, which abounds with ruins, one meets with a dilapidated stone structure consisting of a few cells and a verandah; a door jamb of one of these cells holds in relief a fine image of Padmapānī. Higher up the ground is levelled, and is covered with mounds of bricks, dressed and carved stones, and rubbish. Here two rectangular mounds of brick debris, to the south of the footpath which runs up the hill, represent the ruins of two courts of an ancient monastery, and in the upper one a large statue of Padmapānī stands out prominently. To the north of the ruined monastery are a richly-carved door frame and two images, one of Padmapānī and the other of Tārādevī, which are said to have been exhumed from among the ruins by the *bairāgi* or ascetic of the hill. Still higher up, on the top of the peak, is a circular structure composed of rubble stones, which, according to local belief, was the palace or fort of Rājā Vasukalpa, but which was probably a Buddhist stūpa, like the elevated terrace on the other peak.

On the northern side of the eastern peak is a level plot of ground, 220 feet by 340 feet, said to have been the site of the royal stable, *pilkhānā* and officers' quarters in the time of Rāja Vasukalpa; at the end nearest the hill there are nine statues of stone, almost all of them images of Buddha and Bodhisatwas.

Palmyras Point.—A head-land in the Kendrāpāra sub-division, situated in 20° 46′ N. and 86° 59′ E., which constitutes a landmark for vessels making for the Hooghly from the south.

Ratnāgiri Hill.—A small hill in the Jājpur sub-division, situated 4 miles from Gopālpur on the south bank of the river Kelo in 20° 39′ N. and 86° 20′ E. The flat top of the hill is covered with extensive ruins, discovered by Bābu Manmohan Chakravarti, the most important of which are the remains of the temple of Mahākālī. It faces west and consists of a shrine and hall, the former composed of stones and the latter of brics. The upper portion of the spire of the shrine and the roof of the hall have disappeared, and large trees have taken root in the mouldings of the temple. Near the gate are some fine stone images 1 foot to 3½ feet high, possibly of Tāntrik origin. To the south of the hall is a rough enclosure-wall formed of large blocks of broken statues and mouldings, amongst which are two colossal heads of Buddha. In the centre of the enclosure is a *gumphā* or cave, which a *fakīr* has made by hollowing out a portion of the enclosure. To the west of this enclosure is a rectangular eminence, the centre of which is occupied by a circular mound, evidently the remains of a brick stūpa, with four smaller attendant stūpas at the corners of the enclosure. To the north of the temple of Mahākālī is another rectangular eminence, which probably consists of the ruins of another edifice. To the east of the latter and on a lower level is a row of votive stūpas arranged in a rectangle. Further to the east, statues of Padmapāṇī and other relics, now hidden in jungle, are scattered on all sides. Below this spot to the east of the temple of Mahākālī and the *fakīr's* enclosure, the plateau gently slopes down to another level spot, which is covered with stūpas and statues, among which may be noticed two large statues of Padmapānī. Close by, in a small hollow where the stones and walls still standing presumably represent the remains of an ancient temple, are some seated images of Tārādevī and Padmapānī, more or less mutilated. On the north side of these ruins and about 200 feet north-east of the Mahākālī temple is a large rectangular mass of debris, which, according to local tradition, was a tank where the wives of king Vasukalpa Kesarī used to come to bathe. The water is said to have been brought down from a rivulet to the north, and the villagers still point to some stone pillars as belonging to the aqueduct built by Rājā Vasukalpa. Thick brick walls are traceable on all the four sides of the rectangular mass, inside of which is a hollow of the ancient courtyard. This high rectangular mound in all probability represents the ruins of the largest monastery that once existed here.

Almost all the remains with which the hill is covered are those of religious and not secular monuments, which local tradition ascribes to Vasukalpa Kesarī, the king who is said to have built the palace on the Naltigiri hill. From the abundance of the elaborately-carved images already found there can be little doubt that other remains of great antiquarian interest are still lying buried at the top of the hills. In the enclosure to the east of the temple of Mahākālī is a colossal image with a male figure seated on a lotus and three rows of figures beneath. The head of the image is encircled with a halo from which spring two delicate leafy branches, below which are dots ending in two ducks finely carved. On either side are the figures of armed men mounted on lions couchant on elephants. The whole image has been cut from a solid slab of gneiss and presents a fine specimen of Indian sculpture. Near this are two stones containing Kuthila inscriptions, and two enormous heads of Buddha with thick lips and flat noses of a Dravidian type have also been dug out of the mound on the highest part of the hills. Besides these, stones carved with animals, foliage and arabesque designs are plentiful, and it seems highly probable that excavation would be richly rewarded.

Sārangarh.—At Sārangarh about 5 miles south-west of the town of Cuttack the ruins of an old fort still exist. The word Sārang appears to be a corruption of Churang or Choraganga, the founder of the Gangabansa dynasty (1104—1111 A.D.), the whole name signifying the fort of Churang. The place is now covered with jungle, in the midst of which the ruins of the old fort can still be seen, and close by, a large tank bears the name of Churang *pokhari*. Sārangarh is an important place in the history of Orissa, Kālā Pāhār having fought his last great battle with the chiefs of Orissa under its walls. A detachment of British troops was stationed here for some years after the conquest of Orissa.

Udayagiri Hill.—One of the peaks of the Assia range situated in the Jājpur sub-division in 20° 39′ N. and 86° 15′ E. According to local tradition, the foot of the hill was at one time washed by the sea. It is appropriately named Udayagiri or Sunrise Hill, from its being the most easterly extremity of the Assia range. The chief interest of the place centres in the ruins found on an elevated terrace, sloping from a height of 150 feet down to the level of the plain. This terrace is surrounded, like an amphitheatre, by a semicircle of jagged rocks with an opening towards the east, where it overlooks the Kālia river, flowing about 200 yards from its base.

At the foot of the hill is a colossal statue of Buddha cut in high relief on a single slab of chlorite, which is evidently *in situ*, the lower portion being covered by the debris of an ancient structure. About 200 yards further to the south-west is a large well cut in the rock; it is 23 feet square, 28 feet deep from the top of the rock to the surface of the water, and is surrounded by a stone terrace 94½ feet long and about 39 feet broad. The entrance to the terrace is guarded by two monolithic pillars, the tops of which are broken. The edge of the well and the extremity of the terrace are lined with battlements of large blocks of dressed stone, rounded on the top and 3 feet in height, leaving a wide passage or walk behind. The well is situated at the southern extremity of the terrace. From the north and in the middle of the terrace, a few yards off the entrance, a flight of 31 steps, cut out of the solid rock leads down to the water below. The rock between the lowest step and the well has been cut into an arch, and on its face there is a short inscription in Nāgari, stating that the well belongs to the royal officer Braja Nāga. The same inscription is repeated on the southern parapet of the flight of steps.

Close beside the enclosure of the well is a small temple built by a *fakīr* out of materials pilfered from the neighbouring ruins; and about 200 feet south of the well is another small low-roofed temple similarly built of odd fragments of old sculptures. Inside have been gathered together some broken images, to which offerings are occasionally made by the villagers; and outside are scattered far and wide numbers of images, votive stūpas and beautifully carved stones. The most remarkable of these is a large four-handed statue of Padmapānī, 7 feet high, with a nimbus round his head, and with seven small niches carved above. On the back of it is a long but mutilated inscription in the Kuthila character in 24 lines. To the south of this image stands an old Buddhist temple which has now fallen into ruins. It faces north, and consists of a shrine, hall and porch. In the interior of the sanctum is a large image of Buddha in a sitting and meditative posture. In 1870 the image could be seen whole and entire, but it is now buried up to the breast in the debris of the shrine, while in front lies the fallen lintel of the doorway of the shrine. The hall also is now in ruins, but it was standing as late as 1870 and was supported by rectangular pillars. The door frame of the hall had then on each side four series of elaborate carvings, which have now disappeared.

A little to the west of this temple is a hillock, the top of which is strewn with bricks and rubble, the remains probably of some stūpas. Still further west are to be seen some extensive ruins, of which the most prominent are those of a brick stūpa, the upper portion of which has been destroyed. In the sides facing north and south there are two niches, each containing an image of Buddha, the former of which is inscribed with the well-known Buddhist formula. The other side of the stūpa has similar niches with similar images, but they are now hidden from view, owing to the accumulation of debris. To the south of the stūpa is a large mound now hidden in jungle which is covered with the ruins of what seems to be a Buddhist monastery ; and about 150 feet to the north-west of the stūpa are the rectangular hollows of the foundations of an edifice, from which materials are said to have been carried off by the *fakīr* mentioned above for the construction of new temples.

INDEX.

A.

B.

C.

L.

M.

T.

B. S. Press—10302—520—2-8-1906—E. J. H. & W. M. D'C.